Urban Sustainability

The Urban Sustainability Book Series is a valuable resource for sustainability and urban-related education and research. It offers an inter-disciplinary platform covering all four areas of practice, policy, education, research, and their nexus. The publications in this series are related to critical areas of sustainability, urban studies, planning, and urban geography.

This book series aims to put together cutting-edge research findings linked to the overarching field of urban sustainability. The scope and nature of the topic are broad and interdisciplinary and bring together various associated disciplines from sustainable development, environmental sciences, urbanism, etc. With many advanced research findings in the field, there is a need to put together various discussions and contributions on specific sustainability fields, covering a good range of topics on sustainable development, sustainable urbanism, and urban sustainability. Despite the broad range of issues, we note the importance of practical and policy-oriented directions, extending the literature and directions and pathways towards achieving urban sustainability.

The series will appeal to urbanists, geographers, planners, engineers, architects, governmental authorities, policymakers, researchers of all levels, and to all of those interested in a wide-ranging overview of urban sustainability and its associated fields. The series includes monographs and edited volumes, covering a range of topics under the urban sustainability topic, which can also be used for teaching materials.

Eugenio Mangi · Weixuan Chen · Tim Heath ·
Ali Cheshmehzangi
Editors

Innovative Public Participation Practices for Sustainable Urban Regeneration

Springer

Editors
Eugenio Mangi
Department of Architecture and Built Environment
University of Nottingham, Ningbo Campus
Ningbo, China

Tim Heath
Department of Architecture and Built Environment
University of Nottingham
Nottingham, UK

Weixuan Chen
School of Architecture
The Chinese University of Hong Kong
Hong Kong SAR, China

Ali Cheshmehzangi
Center of Innovation for Education and Research (CIER)
Qingdao City University
Qingdao, Shandong, China

School of Architecture, Design and Planning
The University of Queensland
Brisbane, Australia

ISSN 2731-6483 ISSN 2731-6491 (electronic)
Urban Sustainability
ISBN 978-981-99-9594-3 ISBN 978-981-99-9595-0 (eBook)
https://doi.org/10.1007/978-981-99-9595-0

This Springer imprint is published by the registered company Springer Nature Singapore Pte Ltd.
The registered company address is: 152 Beach Road, #21-01/04 Gateway East, Singapore 189721, Singapore

We collectively dedicate this book to urban activists and scholars around the world who believe in the true value of public participation in urban planning and design processes.

Acknowledgements

We would like to recognize the hard work of all authors and contributors to this exciting book project. We acknowledge their dedication during the whole process, ensuring this book is a success. We also acknowledge the full support of our publishing team at Springer Nature. Ali Cheshmehzangi acknowledges the invaluable support from his external collaborators in Australia, China, Japan, and the UK.

About This Book

In the contemporary processes of urban development, regeneration is becoming a relevant way to limit the occupation of new land and to enhance the existing spatial, social, and ecological dimensions of specific parts of the city. It generally entails the engagement of different groups of stakeholders and users at various levels—e.g., institutional and private—who carry on their own agendas while conveying a wide range of interests and values to safeguard. At the same time, despite indications of a significant increase in its implementation under various forms and in multiple contexts in the last 20 years, participatory mechanisms have not been without challenges and barriers due to several factors, including facilitators' poor negotiation and communication practices, and consequently misinterpretation of the values at stake made by the parties involved, or obstructionism and gridlocks carried on by the different stakeholders.

The narrative of the collection is structured in four sections to critically address and analyze the adoption of stakeholders' engagement in various fields and scales while providing the readers with an international outlook of the topic. The contributions elaborate on innovative participatory patterns, how they are/might be entangled with the different stages of regeneration projects, in what measure they have contributed to the United Nations sustainability goals, and the potential matters emerging during the negotiation phases. In detail, the four parts are:

1. *Participation in Public Space Regeneration*
2. *Participation in Built Heritage Regeneration*
3. *Participation in Informal Settlements Regeneration*
4. *Participation in Ecological Restoration*

Finally, this volume of global case study examples is a significant asset to urban policymakers, planners, practitioners, and researchers in the architecture and urban studies domain who aim at deepening public participation practices knowledge by comparing different experiences.

Eugenio Mangi
Weixuan Chen
Tim Heath
Ali Cheshmehzangi

Contents

Editors and Contributors

About the Editors

Eugenio Mangi is an Assistant Professor in Architecture in the Department of Architecture and Built Environment at the University of Nottingham, Ningbo, China, and he has been an adjunct professor at the Master Program of the College of Design and Innovation, Tongji University. Prior to joining UNNC, he has developed, first with international practices, and then with his own firm, a large number of projects ranging from interior to urban design and heritage renovation in China and Europe. His research interests focus on sustainable urban transformation process in China, local community engagement and participation, urban morphology, and heritage re-use.

Weixuan Chen is a Postdoctoral Fellow at the School of Architecture, The Chinese University of Hong Kong. With a Ph.D. in Built Environment from the University of Nottingham, his research keenly focuses on urban regeneration and governance within the Chinese context. Prior to arriving at University of Nottingham, he completed a bachelor's and a master's Degree from South China Agricultural University and Queen's University Belfast and participated in 10+ planning projects in China as a planner. His contributions to the academic community are highlighted by his work published in respected journals such as Political Geography and Habitat International.

Tim Heath is a Professor of architecture and urban design at the University of Nottingham. He is a qualified architect, town planner, and urban design practitioner having worked extensively in the UK and China where he has undertaken consultancy and projects related to architecture, urban design, and building and urban environmental performance assessments. He is the Course Director of the M.Arch. Sustainable Urban Design and the Director of Postgraduate Programmes. He has previously been the Director of the Institute of Architecture, Head of the School of the Built Environment, and subsequently, Head of the Department of Architecture and Built Environment. He has also been the Associate Dean for Internationalisation and External Relations in the Faculty of Engineering and Acting Vice Provost for Research and Knowledge Exchange at the University of Nottingham, Ningbo, China. He has also acted as an external examiner in many prestigious universities around the World for undergraduate and postgraduate programmes as well as being a Ph.D. examiner. He has chaired many academic and professional accreditation and validation panels in the UK, Malaysia, China, and the UAE's Commission for Academic Accreditation. He is also the author of over 200 academic papers and many books including *Public Places Urban Spaces* (2010); *Revitalising Historic Urban Quarters* (2013); and *Heritage-Led Urban Regeneration in China* (2017).

Ali Cheshmehzangi is the World's top 2% field leader, recognized by Stanford University. He is Professor of Architecture and Urban Planning as well as Head and Founding Director of the Center of Innovation for Education and Research (CIER) at Qingdao City University. He sits in the University's senior management, leading international communications and global partnership inititiaves between China and the UK. Previously, he was a Full Professor in Architecture and Urban Design, Head of the Department of Architecture and Built Environment, Founding Director of the Urban Innovation Lab (UIL), Director of Center for Sustainable Energy Technologies, Head of Research Group for Sustainable Built Environment, and Interim Director of Digital Design Lab. He is the newly-appointed Head of School of Architecture, Design and Planning (ADP) at the University of Queensland, Birsbane, Australia. He was a specially-appointed Full Professor and is now a Research Associate of the Network for Education and Research on Peace and Sustainability (NERPS), Hiroshima University, Japan. He has collaborated and worked with major international and national organizations such as the Asian Development Bank (ADB), the UN Habitat, The Guangzhou Award, the National Development and Reform Commission (NDRC), China Academy of Building Research (CABR), Building Research Establishment (BRE), etc. So far, he has published over 450 journal papers, articles, conference papers, book chapters, and reports. He also has 25 other academic books focused on cities and sustainable development research areas. He has received several awards for three of his books, *Eco Development in China* (2018), *The City in Need* (2020), and *China's City Cluster Development* (2022). he is globally known for his research on "urban sustainability" and promotes applied research at multiple levels.

Contributors

Miguel Ángel Bartorila School of Engineering, Autonomous University of Queretaro, Santiago de Querétaro, Mexico

Diana M. Benjumea Mejia Centre for Core Learning, College of Interdisciplinary and Experiential Learning, Singapore University of Social Sciences, Singapore, Singapore;
Representation, Communication, and Architecture Research Group, National University of Colombia, Manizales, Colombia

Giorgia Carpi Department of Engineering and Architecture, University of Parma, Parma, Italy

Pengyu Chen Department of Architecture & Urban Planning, Shenzhen University, Shenzhen, China

Weixuan Chen School of Architecture, The Chinese University of Hong Kong, Hong Kong SAR, China

Ali Cheshmehzangi School of Architecture, Design and Planning, The University of Queensland, Brisbane, Australia;
Center of Innovation for Education and Research (CIER), Qingdao City University, Qingdao, China

Dario Costi Department of Engineering and Architecture, University of Parma, Parma, Italy

Lingling Deng School of Computer Science, University of Nottingham Ningbo, Ningbo, China

Andrea Fanfoni Department of Engineering and Architecture, University of Parma, Parma, Italy

Md. Nawrose Fatemi Department of Architecture, University of Asia Pacific, Dhaka, Bangladesh

Tim Heath Department of Architecture and Built Environment, University of Nottingham, Nottingham, UK

Zareen Habiba Islam Department of Architecture, University of Asia Pacific, Dhaka, Bangladesh

Jiahui Ji Department of Architecture and Built Environment, University of Nottingham, Nottingham, UK

Luca Lanzoni Department of Architecture and Building Technology, Technical University of Cartagena, Cartagena, Spain

Eugenio Mangi Department of Architecture and Built Environment, University of Nottingham, Ningbo Campus, Ningbo, China

María Jesús Peñalver Martínez Department of Architecture and Building Technology, Technical University of Cartagena, Cartagena, Spain

Emanuele Ortolan Department of Engineering and Architecture, University of Parma, Parma, Italy

Tahmina Rahman Department of Architecture, University of Asia Pacific, Dhaka, Bangladesh

Surabhi Ratavjia School of Computer Science, University of Hertfordshire, Hertfordshire, UK

Prapa Rattadilok School of Computer Science, University of Hertfordshire, Hertfordshire, UK

Chapter 1
Prioritizing Sustainable Urban Regeneration Practices: Addressing Contemporary Urban Challenges Through the Lens of Public Participation

Eugenio Mangi, Weixuan Chen, Tim Heath, and Ali Cheshmehzangi

1 A Brief Introduction

The recent history of urbanisation delineates a trajectory marked by rapid transformations. As urban fabric expands, cities are facing unprecedented pressure caused by the exponential growth of the population, leading to the typical phenomena that affect the built environment, such as land consumption and degradation, urban sprawl, and socio-economic disparities. Meanwhile, the modification of the nature of the land deeply affects biodiversity and alters ecosystem productivity through loss of habitat, biomass, and carbon storage (Seto et al., 2012). Since more than 50% of forcibly displaced people live in urban areas (The World Bank, 2023), rising conflicts among different social groups contribute to further pressure on the urban dimension. The end results are often urban challenges that could be addressed from various sustainable development perspectives (Cheshmehzangi et al., 2021). Nonetheless, many of these pressures continue to last long and become challenges that require multiple stakeholder intervention.

E. Mangi
Department of Architecture and Built Environment, University of Nottingham, Ningbo Campus, Ningbo, China

W. Chen
School of Architecture, The Chinese University of Hong Kong, Hong Kong SAR, China

T. Heath
Department of Architecture and Built Environment, University of Nottingham, Nottingham, UK

A. Cheshmehzangi (✉)
Center of Innovation for Education and Research (CIER), Qingdao City University, Qingdao, China
e-mail: Ali.Chesh@qdc.edu.cn

School of Architecture, Design and Planning, The University of Queensland, Brisbane, Australia

E. Mangi et al. (eds.), *Innovative Public Participation Practices for Sustainable Urban Regeneration*, Urban Sustainability, https://doi.org/10.1007/978-981-99-9595-0_1

Suppose we focus on the cities across the "Global South". In that case, it is possible to highlight that while they are embracing unprecedented (and unavoidable) urban development, they are also confronted with growing pressure on infrastructure, rising costs of service provision, and deepening spatial inequities. At the same time, governments at all levels and residents are called to deal with heavy economic and environmental burdens (Cheshmehzangi, 2023; Liu et al., 2022; Zuk et al., 2018), again putting further pressure on social and spatial dimensions in cities and urban environments.

While urban regeneration is a practice that has been broadly adopted since the post-World War II period in the European contexts, it is only in the last two decades that it has emerged as a strategic response to address these concerns, aiming to harmonise development and conservation. According to the United Nations' report *Urban Regeneration as a Tool for Inclusive and Sustainable Recovery* (2021) this practice should be an integrative process and an essential instrument professionals and policymakers should adopt, while engaging with civil society, to achieve inclusive and sustainable recovery. Its primary goal is to achieve sustainability, inclusivity, and comprehensive urban development without additional environmental burdens (United Nations, 2021). Finally, sustainable urban regeneration should become the preferred practice in consolidated contexts to govern change, attract investments, engage stakeholders (Cheshmehzangi & Dawodu, 2019), and enhance environmental sustainability and climate change resilience (Cheshmehzangi et al., 2023) in both developed and developing countries (European Environment Agency, 2019).

Starting from the previous considerations, it can be affirmed that, in the current urban development frameworks, the role of regeneration has become an essential practice in addressing emerging urban matters. Since the regeneration practices are intended to modify and improve existing and consolidated, integrating the various stakeholders becomes a crucial factor in determining the failure or success of the entire process. Notably, while there has been an increased adoption of participatory mechanisms in urban regeneration over the past two decades, several challenges still persist and need to be addressed. Miscommunication, divergent interests, and stakeholder disagreements complicate the regeneration landscape, which are seen in many different modes of urban development, and in particular, in urban regeneration practices.

While there exists extensive scholarship delving into urban regeneration practices, much of it highlights successful regeneration efforts in various cities, emphasising the role of public spaces, heritage sites, and infrastructure upgrades (Cheshmehzangi & Willman, 2023; Oevermann et al., 2015; Pan & Du, 2021; Wang et al., 2022; Zhang et al., 2022). A consistent theme in these works is the relevance given to the community and stakeholder engagement. As Li et al. (2020) have noted, public participation is essential to ensure that regeneration efforts reflect the community's needs and aspirations. Such engagement also aligns with the sustainability goals set by the United Nations, emphasising shared responsibility and inclusivity (United Nations, 2021). Yet, challenges remain. Despite the acknowledged importance of participation, there are consistent barriers, such as differing stakeholder priorities and bureaucratic impediments (Chen et al., 2022, 2023; Yu et al., 2021). Currently, few studies thoroughly investigate the innovative methods of public participation and their direct influence on sustainable urban regeneration. A deeper understanding of these methods and their implications is required, especially in varied global contexts. Thus, this book delves into this critical matter from the lens of innovative public participation practices in achieving sustainable urban regeneration.

2 Aim and Objectives of the Book

This book is structured into four parts, each focusing on the relation between stakeholders' engagement and urban regeneration practices. The approach is global, examining various contexts and scales (Fig. 1). Specifically, these four parts cover:

Fig. 1 Summary of the book's four parts

1. Participation in public space regeneration
2. Participation in built heritage regeneration
3. Participation in informal settlements regeneration
4. Participation in ecological regeneration.

Drawing from global case studies, this edited volume introduces readers to the implementation of a wide range of participatory practices and how they are applied to different contexts, exploring their efficacy, challenges, and shortcomings. Targeted at urban planners, architects, and researchers, this book not only details existing knowledge but also provides practical insights derived from various urban contexts. The overarching goal is to foster understanding and promote effective practices in public participation for sustainable urban regeneration.

3 Structure of the Book

The book is a collection of eight case study chapters from global perspectives and practices. Each part includes two case studies highlighting specific innovative public participation practices for sustainable urban regeneration. Hence, a global outlook is provided to cover a comprehensive range of different methods, policies, practices, and potential paradigms.

Part 01: Participation in Public Space Regeneration

The first part introduces case studies in Italy and China, addressing the mechanisms of public participation practices aimed at regenerating public spaces.

In their chapter, "*A model of innovative participation in urban regeneration processes of venues born from a collective vocation: Parma città d'oro experience*", the authors Dario Costi, Giorgia Carpi, Andrea Fanfoni and Emanuele Ortolan introduce how public participation aimed at the regeneration public spaces is an emerging practice that involves civil society and institutional actors in Italy. However, the recent experiences leave some perplexities regarding their effectiveness, suggesting the necessity of codification. The end of massive urban development caused by the 2008 financial crisis marked the beginning of the widespread adoption of urban regeneration practices, also encouraged by the policies implemented in the *Zero Land Consumption* season. During this phase, the focus of urban development became the existing city as the only possible context for intervention.

In this scenario, the *Urban Strategic Design* (*USD*) tool proposed by the authors integrates the enhancement of the urban identities, and their attractiveness, with the design of a cycle-pedestrian mobility infrastructure that connects public spaces, urban parks and the historic city center. This methodology provides the institutional actors with a tool to operate on the consolidated urban fabric and the possibility to anticipate scenarios and share them in an oriented participation process.

The USD methodology, applied for the first time in the Italian city of Parma, finds its ultimate validation through the urban design project and a series of events aimed at enhancing the public participation process and experts' involvement. The first opportunity was, in 2016, the exhibition of *Parma Città Futura* (Parma Future City), combined with a participatory process of 10 roundtables with experts on the city's issues. Five years later, in 2021, the initiative *Parma città d'oro* (Parma Golden City) represented a further validation of the experimental tool and its application on the city. The update of the urban projects was the basis for a participatory process articulated on two converging levels: the discussion with experts and the involvement of civil society. The public discussion adopted an innovative tool, the *Civil Atlas of Architecture*, which represents a census of Parma's historical heritage born from a collective vocation. From this participatory design process emerged the importance of integrating two levels of involvement: a top-down participation with the experts, who offered guiding contributions, and a bottom-up participation with the population.

In the second chapter of this part, Lingling Deng et al. focus on *Implementing a participatory design approach to create a sensory-friendly public space for children with special needs* present a participatory design approach for involving children with special needs in public space regeneration. Specifically, this chapter focuses on children with neurodevelopmental disabilities who are hyper- or hypo-sensitive to sensory stimuli or have a unique sensory interest. The study aims at employing the participatory design framework to build a sensory-friendly outdoor space in a rehabilitation centre that caters to around 300 children with Autism Spectrum Disorders (ASD) in the city of Ningbo, China. The participatory design frame adopted in this study is based on online questionnaires in which parents of ASD children reported information about their child's preference when visiting the space. Following the online questionnaires, living lab and focus group methods are organized to involve children as key informants and design partners. The living lab, equipped with virtual reality devices allows children to experience a 'real' regenerated square. The focus groups organized after each living lab session allowed the stakeholders to provide feedback for design iterations. In parallel, the preliminary results obtained from the online questionnaires informed several design considerations for the regeneration of the square. The sensory experience of children at the square will also be supported via sensory stations and smart solutions. This chapter not only provides regeneration recommendations for current outdoor spaces in China, but it also introduces a feasible framework for actively involving children with special needs in the regeneration process.

Part 02: Participation in Built Heritage Regeneration

The second part includes two chapters focused on regeneration for circular cities and a community participation project in the Sanxue historical district in Xi'an, China, addressing the mechanisms of public participation practices aimed at built heritage regeneration.

Luca Lanzoni and María Jesús Peñalver Martínez, in their chapter entitled "*The regeneration of the urban commons: an opportunity for Circular Cities*", introduce the theme of urban regeneration through the lenses of under-utilized spaces. In fact, the majority of contemporary cities often encompass buildings and open spaces, both public and private, which are dismissed, empty, under-used, or abandoned due to several reasons. These assets might be located in the most diverse areas of the urban fabric, such as historical centre, suburban areas, deprived neighbourhoods, etc. The rehabilitation and management of these underused and/or dismissed structures (urban commons) through collaborative processes, is a recognised way to catalyse regeneration and offering social and cultural services with positive side effects in terms of jobs creation, social inclusion, socio-cultural growth, strengthen of local identities, and reduction of urban sprawl.

In this framework, public leverage is essential to promote such collaborative processes, and for this reason, it is important to identify and understand the method to follow and find the appropriate and necessary resources. Despite their relevance, these kinds of practices are still experimental, and there is no systematic application because even if local authorities recognise the added value of such collaborative processes, promoting such practices is not easy for many reasons, mainly:

- Unclear local urban regulation (i.e., legal framework,) to promote the cooperative management of the urban assets (sleeping giants).
- Insufficient human resources and inadequate skills to manage co-creation activities.
- Lack of a design approach to urban regeneration focused on social innovation.

To be able to manage this complexity we must develop regeneration actions, not only through collaborative approaches from the bottom up, capable of making different urban and territorial actors interact but above all also with horizontal actions, capable of making different departments of local authorities in charge for the urban management cooperate, therefore design integrated urban regeneration processes.

Later in Chap. 5 of the book, Jiahui Ji and Tim Heath focus on "*Community participation of China's urban regeneration: a case study of Sanxue Historical District in Xi'an*". In this chapter, the authors analyse the participatory practices that were adopted to regenerate a historical district in Xi'an, China. According to them, in the past two decades, community participation has been playing a critical role in urban regeneration. Due to the persistent top-down tendency, particularly in developing countries, expanding the role of community participation in urban regeneration practices is vital. As a response to the criticism of high-centralised regeneration projects, thus achieving a more inclusive approach and enhancing positive interactions among different stakeholders, Chinese communities are increasingly empowered in the decision-making process. Through an appropriate participatory method, people could have an opportunity to act in the regeneration process, and cooperate with other stakeholders to complete sound, long-term, and local-based strategies. However, although the community participation approach has been launched, people's demands have failed to be satisfied due to the continuing marginalising of the community. In order to dissect the reasons, by taking the Sanxue historical district in Xi'an as a case study, this paper aims to elaborate on a new community participation model–collaborative workshop; analyses how collaborative workshop operates and why it finally results in the ineffective of participation. Grounded in a qualitative research methodology, this paper is built upon data collected through archival research, interviews with the local community as well as other stakeholders, and on-site observation.

Part 03: Participation in Informal Settlements Regeneration

The third part includes two chapters from Bangladesh and China, addressing the improvement projects for informal settlements regeneration.

Chapter 6 titled "*Improving liveability through regenerating informal neighborhoods: a case of Old Dhaka, Bangladesh*", authored by Md. Nawrose Fatemi et al. delves into the case of neighbourhood improvement. The authors discuss the City of Dhaka, Bangladesh, has been the world's most densely inhabited city since 2016. It has a density of 30,093 people/km^2 in 2022, posing challenges in providing access to basic civic facilities to its residents. The planning history of Dhaka reveals that both planned and spontaneously developed neighbourhoods have evolved throughout time. Besides these two distinct types of urban developments, the historic part of Dhaka, with its cultural and social identity has expanded within its physical limits. Consequently, citizens' liveability conditions in Old Dhaka characterized by the low standard of urban services, narrow road networks, poor drainage systems and congested urban fabric, have steadily worsened over time. A new Detailed Area Plan (DAP) has recently been restructured and published to develop Dhaka into a better liveable city (Ruszczyk et al., 2023), resolving its urban challenges, ensuring citizens' equity proportionate to urban amenities, upgrading urban services, and gradually improving the physical, social, economic and environmental condition of different areas. Engaging locals during the fieldwork, this study attempts to identify the existing shortcomings, their experience living with these challenges and the aspirations of the locality in relation to the regeneration proposal of DAP. Several participatory approaches were exercised, including focus group discussions, key informant interviews, transect walks and community mapping of the existing conditions and two urban design workshops were conducted with 50 participants to demonstrate the urban regeneration proposals of DAP for two Old Dhaka neighbourhoods. This paper suggests solutions in terms of policy and design to provide a liveable environment for the locals in terms of physical infrastructure, amenities and facilities while conserving the socio-cultural fabric of the neighbourhoods. This study concludes that balancing top-down and bottom-up approaches in design solutions can ensure liveability not only in terms of physical outcomes but also regenerate the embedded socio-cultural aspects of these neighbourhoods.

In the following chapter by Pengyu Chen and Tim Heath, the authors explore "*The New Strategy of Urban Village Regeneration: the comprehensive improvement project in Jingle Village, Shenzhen, China*". They argue that urban villages are a common yet heterogeneous informal living space in China's developed cities, characterised by high density, a high degree of autonomy, and difficulty in being regulated by the government. As a provider of low-cost residential accommodation, they have contributed significantly to the successful growth of cities. However, they have also posed a number of problems, such as hygiene and security. During rapid industrialisation and urbanisation in China, the government has often considered demolition and redevelopment a convenient and cost-effective way of regenerating urban villages. However, with the recent shortage in the supply of land resources and soaring land prices, the demolition and redevelopment approach has created a series of social and policy issues.

Recently, comprehensive improvement—a way of beautifying the built environment and optimising management without altering the building construction—has been introduced as an alternative approach to regenerating urban villages, and it is anticipated as a sustainable model of urban regeneration. The planning and implementation process involves multiple stakeholders, including the government, developers, villagers, tenants, etc. These different actors, however, all have different interests and expectations in the processes of planning, implementation, and operation. While contributing to the sustainability of the regeneration projects, many potential problems have also arisen, some of which have led to small-scale open protests.

This chapter uses Jingle Village in Shenzhen as a case study to examine the interests of different stakeholders at different stages of the comprehensive improvement process and the problems faced in implementing the project. The chapter critically assesses and analyses the positions and roles of the various stakeholders at different levels based on in-depth semi-structured interviews and resident questionnaires to assess the sustainability of the new regeneration model. The findings can contribute significantly to the sustainability of urban regeneration strategies in China and even the global South. The analysis of the different levels of stakeholders also will be of great value in informing many different types of regeneration policies.

Part 04: Participation in Ecological Regeneration

The fourth and last part of the book includes two chapters from Latin America, Colombia and Mexico, providing methods and practices for recovering the ecological dimension of degraded urban settings.

The first case of a Latin American city is introduced by Diana M. Benjumea Mejia, titled "*An Ecology Model for Participatory Strategies: Community-led Green Networks and its Social and Spatial Agents*". The city of Medellin has sustained dramatic changes over the past three decades, from standing as one of the most violent cities in the 1990s to becoming one of Colombia's most resilient, innovative, and greenest cities in recent times. A key component of these changes was the transformation of the lower-income neighbourhoods by implementing participatory strategies to reduce violence and increase social resilience and inclusion through urban green revitalisation projects. A remarkable initiative was the Green Belt project included in the Plan Maestro Area Centroriental, designed to connect the most relevant nodes through walking and cycling paths, leisure centres, and urban green infrastructure (e.g., urban farming, community gardens, parks). The Government-led participatory strategies have beautified the neighbourhoods and attracted local and international attention. Nevertheless, the shortcoming of these strategies in Medellin is the fragile relationship between stakeholders and space ownership flounders over the green spaces. The conflicts occur when green spaces, which were previously managed by the communities, are transferred to the local governments, fracturing both activism and spatial equality. An overlooked green infrastructure that has survived for decades is the small-scale community-led green interventions often located at the fringes of residual green areas (i.e., nearby roads or areas next to water streams). The random appropriation of residual green spaces depicts vital signs of pro-environmental behaviours, including nature preservation and environmental agency as well as a green network that acknowledges the participation of both human and non-human living agents (e.g., plants, trees, and wildlife)—often overlooked in urban design and planning. This chapter discusses the potential of these small-scale community-led green networks as complementary paths to government participatory strategies in urban green revitalisation projects. Different living agents that intervene in the process, the levels of interventions, and forms of autonomous participation are unraveled and analysed to formulate an ecology of participation model that introduces a paradigm of equitable interactions in urban green projects based on a living ecology framework that acknowledges the social and political agency of all spatial actors.

In the last chapter of the book, titled "Cultural and natural heritage, synergy in regeneration of tropical cities: Laguna del Carpintero wetland and its surroundings, Gulf of Mexico", Miguel Ángel Bartorila stresses how Latin American cities, with their incomplete urban development processes and marginalised natural spaces, adopt regeneration practices to recover and improve the existing spatial, social, and ecological dimensions. Since it is assumed that natural ecosystems are a key element in sustainable urban regeneration projects, this chapter aims to explore the synergy between natural and urban processes in urban renovation projects through the integration of wetland areas, particularly mangrove areas.

This text dissects the Conservation and Revitalisation Master Plan for the Laguna del Carpintero wetland and its surroundings located in the port of Tampico, Mexico. The project focuses on the integration of urban biodiversity around open spaces. The author introduces the innovative methodological aspects adopted in the project, which is located in a tropical city context. It conceptually develops the essential material components for understanding and formulating projects for Sustainable Urban Regeneration, including ecological succession. Therefore, mutual adaptation between urban and natural components associates ecological restoration with public space and built heritage in the project logic.

4 Conclusions

As urbanisation continues to reshape the global landscape, the challenges it presents—from environmental impact and biodiversity loss to the socio-economic pressures linked to residents' displacement and rapid expansion—become increasingly relevant. This trajectory of rapid transformation necessitates a strategic and holistic response, and urban regeneration can play a pivotal road in this sense. Aiming to weave the threads of conservation and development, urban regeneration proposes an integrative path to sustainability, inclusivity, and comprehensive growth without augmenting environmental pressures.

The framework of this book is meticulously designed to offer a panoramic view of the urban regeneration landscape, zooming into its nuances through three primary lenses: the regeneration of public spaces and ecosystems, the preservation and rejuvenation of built heritage, and the restructuring of informal settlements. By offering this tripartite approach, the book strives to provide an encompassing understanding of the myriad facets of urban regeneration, anchored by global case studies. These cases not only present a testament to the successes and challenges of the past but also provide a roadmap for the future.

It becomes apparent that its primary mission is twofold: first, to provide a comprehensive overview of the state of urban regeneration across different global contexts, and second, to underscore the paramount importance of effective stakeholder engagement. Through its chapters, the book not only enriches the academic discourse around (sustainable) urban regeneration but also offers pragmatic insights for professionals on the ground. These will be looked at more precisely from the lens of public participation in urban regeneration practices.

References

Chen, W., Cheshmehzangi, A., Mangi, E., Heath, T., Ye, C., & Wang, L. (2022). An analysis of residents' social profiles influencing their participation in community micro-regeneration projects in China: A case study of Yongtai Community, Guangzhou. *Land, 11*.

Chen, W., Ye, C., & Liu, Y. (2023). From the arrival cities to affordable cities in China: Seeing through the practices of rural migrants' participation in Guangzhou's urban village regeneration. *Habitat International, 138*.

Cheshmehzangi, A. (2023). *Mapping urban regeneration: City life experiences in Yunnan*. Springer.

Cheshmehzangi, A., & Dawodu, A. (2019). *Sustainable urban development in the age of climate change—people: The cure or curse*. Palgrave Macmillan.

Cheshmehzangi, A., Dawodu, A., & Sharifi, A. (2021). *Sustainable urbanism in China*. Routledge.

Cheshmehzangi, A., He, B., Sharifi, A., & Matzarakis, A. (Eds.). (2023). *Climate change and cooling cities*. Springer.

Cheshmehzangi, A., & Willman, E. (2023). The city as palimpsest to redefine the role of architectural continuity in re-establishing urban identity in post-industrial landscapes. *Journal of Urban Regeneration and Renewal, 16*(3) (Spring 2023).

European Environment Agency. (2019). *Sustainability transitions: Policy and practice*.

Li, X., Zhang, F., Hui, E. C.-M., & Lang, W. (2020). Collaborative workshop and community participation: A new approach to urban regeneration in China. *Cities, 102*.

Liu, S., Zhang, F., & Wu, F. (2022). Contrasting migrants' sense of belonging to the city in selected peri-urban neighbourhoods in Beijing. *Cities, 120*.

Oevermann, H., Degenkolb, J., Diessle, A., Karge, S., & Peltz, U. (2015). Participation in the reuse of industrial heritage sites: The case of Oberschöneweide, Berlin. *International Journal of Heritage Studies, 22*, 43–58.

Pan, W., & Du, J. (2021). Towards sustainable urban transition: A critical review of strategies and policies of urban village renewal in Shenzhen, China. *Land Use Policy, 111*.

Ruszczyk, H. A., Halligey, A., Rahman, M. F., & Ahmed, I. (2023). Liveability and vitality: An exploration of small cities in Bangladesh. *Cities, 133*.

Seto, K. C., Güneralp, B., & Hutyra, L. R. (2012). Global forecasts of urban expansion to 2030 and direct impacts on biodiversity and carbon pools. *Proceedings of the National Academy of Sciences, 109*, 16083–16088.

The World Bank. (2023). *Urban development* [Online]. The World Bank. Retrieved September 13, 2023, from https://www.worldbank.org/en/topic/urbandevelopment/overview

United Nations. (2021). *Urban regeneration as a tool for inclusive and sustainable recovery*. United Nations Human Settlements Programme Office.
Wang, M., Zhang, F., & Wu, F. (2022). Governing urban redevelopment: A case study of Yongqingfang in Guangzhou, China. *Cities,* 120.
Yu, Y., Hamnett, C., Ye, Y., & Guo, W. (2021). Top-down intergovernmental relations and power-building from below in China's urban redevelopment: An urban political order perspective. *Land Use Policy, 109.*
Zhang, H., Cong, C., & Chakraborty, A. (2022). Exploring the institutional dilemma and governance transformation in China's urban regeneration: Based on the case of Shanghai Old Town. *Cities, 131.*
Zuk, M., Bierbaum, A. H., Chapple, K., Gorska, K., & Loukaitou-Sideris, A. (2018). Gentrification, displacement, and the role of public investment. *Journal of Planning Literature, 33*, 31–44.

Part I
Participation in Public Space Regeneration

Chapter 2
A Model of Innovative Participation in Urban Regeneration Processes of Venues Born from a Collective Vocation: Parma Città d'Oro Experience

Dario Costi, Giorgia Carpi, Andrea Fanfoni, and Emanuele Ortolan

Abstract In Italy, participation is an emerging practice in the urban regeneration processes of public spaces. However, the experiences of the last few years leave some perplexities regarding their effectiveness, suggesting the necessity of codification. The end of urban sprawl led cities into a crisis scenario marking the beginning of the urban regeneration and *Zero Land Consumption* season. During this phase, the focus shifts from full spaces to empty ones and the existing city becomes the only possible context for intervention. In this scenario, the *Urban Strategic Design* (*USD*) tool integrates the enhancement in terms of urban identities attractiveness with the design of a cycle-pedestrian mobility network, leading to public spaces, urban parks and the historic center. This methodology provides to public authorities a support to operate on the settlement with the possibility to foresee scenarios to be shared through an oriented participation process. The *USD* methodology finds its ultimate validation in the project and the journey of sharing for the city of Parma. The first opportunity was, in 2016, the exhibition of *Parma Città Futura* (Parma Future City), combined with a participatory process of 10 roundtables with experts on the city's issues. Five years later, in 2021, the initiative *Parma città d'oro* (Parma golden city) represented a further validation of the experimental model on the city. The update of the urban projects was the basis for a participatory process articulated on two converging levels: the discussion with experts and the involvement of the citizens. The public discussion adopted an innovative tool, the *Civil Atlas of Architecture*, which represents a census of Parma's historical heritage born from a collective vocation. From this participatory design process emerged the importance of integrating two levels of involvement: a top-down participation with the experts, who offered guiding contributions, and a bottom-up participation with the population.

D. Costi · G. Carpi (✉) · A. Fanfoni · E. Ortolan
Department of Engineering and Architecture, University of Parma, Parma, Italy
e-mail: giorgia.carpi@unipr.it

E. Mangi et al. (eds.), *Innovative Public Participation Practices for Sustainable Urban Regeneration*, Urban Sustainability, https://doi.org/10.1007/978-981-99-9595-0_2

1 Introduction

The participation in urban regeneration processes of public spaces in Italy is becoming a widespread practice, thanks to the presence of national and regional legislation that guarantees its application.[1]

However, the design and regeneration of public spaces commonly adopted only offers the illusion to non-institutional actors of being able to influence the plan choices, because their participation usually occurs subsequently, when the decisions have already been taken. Whenever the population is involved at the beginning of the decision-making process, information and communication methods are often unclear, incomplete and partial (Pontradolfi, 2002). Participation has in fact become an aim in itself of interventions to transform urban space, rather than a tool for listening and involvement.

The tool of participation should become a technical requirement of territorial government. On the other hand, it must to assume a strategic role for the quality of projects, leading to the development of living environments, capable of expressing the characteristics of the place. Participatory design is in fact based on architecture's capability to recognize and enhance the *identity of places* and to generate awareness in citizens, through design choices shared by the community thanks to the involvement of users at all stages of the process. Giancarlo De Carlo, one of the main protagonists of participatory design practice in Italy, has focused his research on *urban identity* and the need to recognize and preserve it (De Carlo, 1995). If the loss of the specific identity of places in the city makes all forms of communication and confrontation impossible, participatory practice becomes the tool to bring people, back to self-awareness through an identity acknowledgement of places (Mazzoleni & Sepe, 2005).

The debate on participation is part of a crisis scenario of the disciplines that deal with the design of the city: urban planning, architectural and urban design, which over the last few decades have adopted different, in some ways opposite methodologies of intervention. Urban planning almost never gets to the preliminary check, while architectural design is losing interest in overall issues (Talia, 2011). The crisis scenario that Italian cities are going through underlines the absolute need to recover integration between the different scales of intervention, through a strategic vision shared by the community.

In this context, the *Urban Strategic Design (USD)* tool, developed in recent years, combines two essential working methods: urban design through architectural prefiguration and the sharing of choices through an oriented participation process. This arises as a methodological tool complementary to conventional urban planning. Moreover, it is based on the potential of architectural design to prefigure a scenario of urban perspectives (De Jong, 1992) which allows Public Administrations to draw up an overall strategy of intervention, sharing it clearly with all those involved in the process of city's transformation.

---

[1] On this topic INU (National Institute of Urban Planning) held a conference, 3–6 April 2019, entitled "Mosaic Italy: telling the future".

The *USD* has been refined over time, thanks to continuous field tests. Among these, the most in-depth is the experimentation, begun seven years ago, on the city of Parma, through two main initiatives: *Parma Città Futura* (2016) and *Parma Città d'Oro* (2021).

2 Legislative Framework: European and Italian Policies for Participatory Design

In Europe, since the end of the twentieth century, there has been a radical transformation in the social structure of cities and a growing awareness of the necessity to take part in the transformation processes of the territory (Robert & Steiner, 1989). This perspective has led the European Union to guide member states to take action according to new criteria and principles of economic and social cohesion, sustainable development and balanced competitiveness (Carta, 2003). The main policies related to citizenship involvement in territorial regeneration interventions are:

- *The Agenda XXI Program* (Rio de Janeiro, 1992) states: "the result of a combination of integrated policies and the active participation of local community actors".[2]
- *The Aarhus Convention* on "*access to information, public participation and access to justice*" (1998), which demanded that governments effectively inform citizens when starting a decision-making process related to environment.[3]
- *The European Charter of Human Rights in the City* (*Saint Denis Charter*, 2000), which is an instrument used by institutions to implement policies closer to the citizen.[4]
- *The White Paper: European Governance* (2001), by which participation became the subject of specific political action.[5]
- In 2002, with the document: '*Towards a culture of greater consultation and dialogue. General principles and minimum standards for consultation of interested parties by the Commission*', the EU emphasizes that participation, openness and consistency are the general principles guiding consultation.[6]

[2] Bollini G, Bologna G, Calori A, Merola M, Un'introduzione all'Agenda 21 Locale: www.flanet.org.

[3] The convention is available at https://unece.org/environment-policy/public-participation.

[4] Article 1 'Right to the city' states: 'The city is a collective space that belongs to all inhabitants, who have the right to find in it the conditions necessary to fulfil their political, social and environmental aspirations.

[5] For further information visit the European Commission's website on governance: http://www.europa.eu.int/comm/governance/.

[6] Available online https://eur-lex.europa.eu/legal-content/IT/TXT/PDF/?uri=CELEX:52002DC0704&from=IT.

- In 2005, the European Commission adopted the *Plan D for Democracy, Dialogue and Debate*, presenting the initiatives to strengthen dialogue, public debate and citizen participation, considered to be the basis for democratic renewal.[7]
- In 2009, the *Treaty of Lisbon* came into force, which enshrined participation as a genuine constitutional principle.[8]

These proposals in Italy are declined through a number of key legislative acts, such as the *Bosetti-Gatti Law* (No. 241/1991) and the *Bassanini Decree* (1997), which helped to promote a new perspective. According to this point of view, individual or associated citizens become allies with whom work on the co-design. These approaches establish new tools for bottom-up economic planning and improve the democratic participation of citizens in the decision-making processes (Pontradolfi, 2002).

In 2014, the National Institute of Urban Planning promoted the signing of the *Participation Charter*, marking the beginning of a path that aims at the progressive involvement of public bodies and associative structures to spread the culture of effective and "quality" participation. The document consists of 10 principles: cooperation, trust, information, inclusion, effectiveness, constructive interaction, equity, harmony, accountability and evaluation, which indicate how to give quality to the participatory process. The Charter can be used as a methodological guide when designing a citizen involvement process, or as an evaluation grid to determine the quality of a proposed participatory process.[9]

In Italy, at the regional level, only Emilia-Romagna, Tuscany and Apulia have drafted a law concerning participation. In detail, Emilia-Romagna has Regional Law No. 3 of 2010, which aims at total participation of the community in the choices that affect the territory. Moreover, it identifies favorable conditions for the assertion of participatory democracy and provides measures to promote and support greater development of active citizenship form (Ciancaglini, 2010). This law will then be replaced by Regional Law No. 15 of 2018,[10] which was designed to introduce substantial changes on the following points:

- Simplified language to achieve greater inclusion;
- Preventive information and participation to citizens on interventions of particular relevance to the community;
- Establishment of the Participation Day;
- Training activities with the aim to promote the culture of participation within the local authorities;

[7] Available online https://eur-lex.europa.eu/legal-content/IT/TXT/PDF/?uri=CELEX:52005DC0494&from=IT.

[8] Available online https://eur-lex.europa.eu/legal-content/IT/TXT/PDF/?uri=OJ:C:2007:306:FULL&from=IT.

[9] https://www.inu.it/la-carta-della-partecipazione/.

[10] https://partecipazione.regione.emilia-romagna.it/la-legge-E-il-bando/legge-regionale-partecipazione/la-legge-regionale-N-15-2018.

– Entrusting the Legislative Assembly with the promotion of the law through seminar and study activities.

3 Literature Review

3.1 The Origins of Participatory Design

The historical and cultural matrix of participatory planning has its roots in the early 1900s with Patrick Geddes, one of the founding fathers of modern town planning and one of the first to address citizen involvement in decision-making processes. He argued that planning was not merely a matter of government between engineers and architects, but was the result of an entire civilization, a community and an epoch. Indeed, he wrote that "the evolution of cities and the evolution of citizens are two processes that must take place together" (Geddes, 1915).

In the same school of thought are the theories of Lewis Mumford. He theorizes an idea of planning as a 'self-educational' process of social transformation. Mumford's theories focus on the relationship between planning and democracy, between political order and an ecological, economic and social equilibrium (Mumford, 1938).

The evolution of citizen involvement theories saw a decisive moment in the second half of the 1960s, with the birth in the USA of the concept of *advocacy planning,* by Paul Davidoff. It was the answer to the US policies of *Urban Renewal* which had led to the loss of a sense of community for many families, who no longer had a say in decisions concerning the transformation of the city. Davidoff stated pluralism in urban planning, known as the possibility for various stakeholders to formulate their own proposals for urban development plans, without necessarily depending on the government's initiative. Indeed, he argued that "when the government has a monopoly on the proposal of plans, debate is limited and the population can only "react" rather than "propose" (Davidoff, 1965).

Another response to the effects of the US *Urban Renewal Act* was *The Ladder of Participation* theorized by Sharry Arnstein in 1969. He provided a clear principle of hierarchization of participatory practices, divided into three levels, corresponding to different registers of practices (Arnstein, 1969) (Fig. 1).

Fig. 1 Ladder of participation (adapted from Arnstein, 1969)

LEVELS OF PARTICIPATION	CORRISPONDING PRACTICES
Effective empowerment of citizen	Citizen control, delegation of partnership power
Symbolic Cooperation (tokenism)	Reassurance(placation), Consultation, Information
None Partecipation	Therapy, Manipulation

At the *Non-Participation* level, participatory practice is aimed exclusively at gaining the support of citizens, through techniques belonging to the sphere of advertising and public relations. The second level of *Symbolic Cooperation* includes three stages: *information, consultation and mediation.* All necessary but not sufficient steps to guarantee an effective participatory process. Only in the third level, *Power to the citizens*, it is possible to speak of a true participation. Also, this level consists of three degrees: *partnership, delegation of power and citizen control.* At the end of 1970s, John Friedmann broke into this context with his theory of *radical planning* as a proposal to the planning crisis (Friedmann, 1987). The thought of Friedmans develops around two key aspects: *knowledge* as an indispensable factor in the governance of urban transformations and *action* (Carta, 1996) as a consequence for the effectiveness on the territory of the prefigured methodologies and analyses.

3.2 Urban Design and Participation in Italy

Since the Second World War, the scientific dissertation of participatory planning and design experiences in the Italian context have been strongly influenced by the translation of Geddes and Mumford's texts. The main protagonists of the new community-based approach in Italy were: Danilo Dolci, Adriano Olivetti, Carlo Doglio, Ludovico Quaroni, Giancarlo De Carlo and Pasquale Culotta. Olivetti, entrepreneur and intellectual, was among the first to realize the importance of the relationship between economy, industry, territory and involvement of the inhabitants (Olivetti, 1952). This attitude is evident in the experience of Matera, for Borgo La Martella, designed by Quaroni, in which the relationships and values present among the inhabitants are preserved through a direct dialogue with the architects (Barbera, 2019). Close to Olivetti was Doglio, who took part in the development of the Ivrea Masterplan (1951–1955), which was the result of close collaboration between technicians and local inhabitants, leading to the elaboration of a wide-ranging plan (Renacco, 1955). Another tangible example of bottom-up participation in those years (1953–1954) is Dolci's Centre for Full Employment in Partinico. The training process of technicians and the community was one of the main goals in the construction phase of the plan (Dolci, 2020).

In the Italian panorama, De Carlo is certainly one of the most authoritative voices on participatory design. In his essay *L'architettura della partecipazione* (The Architecture of Participation), the author dwells on the meaning and implications of participation, which according to him must not be limited to sharing the main choices but should involve the inhabitant in the architectural research, whose the designer is a critical interpreter (De Carlo, 1971). There are two main experiences carried out by De Carlo: the Plan for the historic centre of Rimini and the Matteotti Village in Terni. The participatory practices that marked the Rimini experience (1966–1969) were articulated through a double questionnaire, submitted to inhabitants and tourists. De Carlo focuses on the formal and functional characteristics of the historic center and how these were perceived by citizens (Tomasetti, 2012). Although it has never been realized, the plan for Rimini's historic center represents a fundamental episode in the Italian context, that sanction the importance of participatory processes in the practice of planning.

The project for the workers' residences in the Matteotti Village (1969–1975) has been developed through a dense program of meetings with the population. The architect set the six objectives at the basis of this intervention: the separation of people and cars with appropriate routes, private greenery for each dwelling, collective greenery, neighborhood public services, a building type with few levels and high density, housing types differentiated by household. The meetings and interviews were supported by massive community education through exhibitions and debates to raise awareness of the issues. This operation had the advantage of increasing the housing quality of the built complex and the ability to trigger a process of identification of the inhabitants (Franchini, 2020).

In sync with Giancarlo De Carlo's civil engagement there was also Pasquale Culotta, who in 1987 promoted a collective architectural design applied to the scale of cities, aimed at completing the work done on Palermo the year before for the *Triennale of Milan*. The proposal imagined *Nine Landmarks* to be realized for the National Exhibition planned for 1991, solving a series of strategic places through projects conceived not only as temporary occasions but also for permanent use (Culotta et al., 1988). Despite their different backgrounds and design inclinations, these characters have concept of community relations in common, which is implemented through the involvement of citizens in the design of places.

4 Methodology: *Urban Strategic Design* as an Innovative Experimentation of Public Spaces Design Through a Participatory Approach

The urban crisis, the difficult economic and social context and the recent health crisis have revealed in Italy the incoherence of the last century urban expansion (Kotkin, 2020; Simon, 2020) and the limits of urban governance (Sharifi & Amir Reza, 2020). The halt in the expansion of cities, the start of the *Urban Regeneration* phase and *the Zero Land Consumption* policy provide an extraordinary opportunity to reorder the system of empty spaces and to rethink the existing urban fabric, starting from unbuilt spaces. The periphery urban voids and the fragile landscapes will be the new field of work in the urban regeneration processes (Bertelli, 2018). In this context, the *Urban Strategic Design* (*USD*) methodology integrates existing planning tools with the elaboration of an overall architectural scenario for the contemporary city.

The constitution of a network of public spaces and parks and the reinvention of urban identities transformed into *city ideas*, are the two main objectives behind this methodology. The *Reserch by Design* was adopted as methodological approach because the planning of the city's future cannot be based on program certainties (Roggema, 2017). Through the project, the architect generates an issue and develops a solution in relation to places, contexts and needs (Thomsen & Tamke, 2009). In addition to asserting the central role of the urban project in planning processes, the *USD* uses a coherent working methodology and specific actions that organize participation from the initial design to the final developments in an oriented and effective manner. This favors the construction of an open consensus of the planning choices, shared through the intervention of all the city's actors on the future urban landscape (McCann, 2011).

If the existing city becomes the main context of intervention, the focus shifts from full to empty spaces that constitute the real latent heritage of cities. The inversion of the objective becomes the decisive trigger for the evaluation of urban potentials within the construction of a green, accessible and welcoming city reorganized on a system of places designed for a social mobility alternative to the current road one. The *USD* methodology is articulated through 3 objectives and 12 actions that define *a city idea* shared with citizens to be presented to public institutional stakeholders and private investors.

4.1 Three Goals

The *USD* main goal is to find a way to develop urban identity into *city concept.* Indeed, each urban context has its own uniqueness and character, which finds in some specific buildings and special structures the community identity places (Rossi, 1966). Today, these places can be extraordinary resources for a recovery of awareness, economic revival and territorial promotion.

The second goal considers the real conditions of the city starting from the relationship with its territory. On one hand, the architectural project will have to make use of the environmental system, to be connected to the network of supralocal cycle paths for a new deep-rooted tourism network, and on the other hand, of infrastructural works and hydraulic safety measures as feasible contribution towards the creation of the itineraries. Thus, the watercourses could become the ideal environment for the long and medium-length cycle and pedestrian tracks, reconnecting country roads, as an opportunity to relaunch the tourist vocation of the agricultural context and enhance the value of the farms established there.

The third goal is to realize the *City of People 4.0* as a renewal of the idea of the '*City of Man*', elaborated a few decades ago by Adriano Olivetti in the Season of the Fourth Industrial Revolution, characterized by permanent interconnection and interactivity (Costi, 2021).

4.2 Twelve Actions

The *USD* is an architectural direction tool for the implementation of an overall design of a sustainable and intelligent city, through 12 actions useful to elaborate a strategy for the reorganization of slow urban relations and the definition of a shared *idea of the city*.

- *Action 1—Overall mobilization and an integrated vision.*
  A collective debate spanning different scales and characters, involving a multiplicity of competences.
- *Action 2—A city concept: how a planning community grows together.*
  The enhancement of identities.
- *Action 3—The collective scenario of the smart and sustainable city.*
  The urban application of digital technologies for a Smart City at the service of people.
- *Action 4—The territorial scale: environmental and tourist scenarios for a new nervous system of the urban body.*
  Contextualization of the strategic decisions in the area of reference on an urban scale.
- *Action 5—The power of empty spaces.*
  Systemization of public parks and fragile environments.
- *Action 6—The urban scale: the new nervous system within the fabric for a green unitary system of slow relations.*
  Identification of a set of areas to link across the city to form a new series of cycle-pedestrian routes, separate from the roadway network.
- *Action 7—The locations scale: areas of intervention like nodes of the nervous system in the urban fabric.*
  The choice of places with a collective vocation on which to focus.
- *Action 8—The design as a probe.*
  The architectural design as research and verification into the potential of places.
- *Action 9—Architecture for the city.*
  Architectural prefiguration as a tool for comprehending and sharing the collective challenge. The city concept becomes a strategic scenario to be promoted by participation.
- *Action 10—The integrated digital infrastructure strategy.*
  The urban project as an experiment of the city of the Fourth Industrial Revolution, working at the urban level by designing a continuous system of public spaces and at the underground level by applying innovative Smart Grid technologies.
- *Action 11—The participatory program.*
  Define the modes, times and subjects. Put competences into action (research and design, thematic discussion tables) and mobilize the community (exhibitions, initiatives, competitions).
- *Action 12—Reality and design: self-evaluation and monitoring of the quality of life.*
  Adoption of BES (Benessere Equo Sostenibile—Equitable Sustainable Well-being) to analyse the current situation and assess the future effects of the *USD*.

From these 12 actions, the centrality of the participatory process within the *USD* methodology is evident. The goal is to launch from the earliest stages a structured process of public confrontation that starts with the involvement of knowledge and stakeholders in the area. Three levels of parallel deepening will be necessary: *specific skills, broad skills and experience.*

The first level is the transversal field of *specific competences*, today too fragmented and divided into individual specializations. Different disciplines as diverse as urban planning, environment, mobility, transport, infrastructure, architectural and landscape design will be called upon to confront each other, sharing sectorial strategies to interact and converging under the architectural direction of the urban project.

The second level is the *broad competences* that plays a central role in the elaboration of the *city idea*, through the promotion of identity features, that will have enhanced citizens' critical awareness and urban attractiveness. Experiences are the third component. Civil society, companies, the world of associations, but also parishes and aggregative realities are called upon to take part in the public discussion and make their own contribution. These three levels will be able to converge in thematic in-depth discussions. The transversal and inclusive organization of public discussion tables on issues of city relevance must be coordinated with the objective of developing shared proposals to be offered to the public administration. The matter of the outcome is another point to emphasize, in order to verify the productivity of the process. Too often, these processes are done mainly to communicate a mobilization that in reality proves to be sterile and with no real prospect of implementing the outcomes.

The Public Administration will have two distinct responsibilities: the initial support and the final development. The PA will be able to promote the process while leaving citizens free to develop an independent path, reaping the benefits at the end. In fact, it will be its responsibility to transpose and develop the indications that will emerge and that can be concretely deepened later. The mobilized Civil Society will have to take on a new role: it will be able to demand to be heard, in the knowledge that it is carrying out its essential action of defending Democracy (Cassese, 2017), activating that real process of shared construction of viable scenarios. The primary purpose of this process is the elaboration of concrete proposals. The implementation guidelines are in fact the outcome of the public debate on the future characterization of the city and a fundamental development of the strategic choices made on the architectural and urban plan.

5 Participation in *Urban Strategic Design* for Parma: 7 Years of Experimentation on the City

5.1 Parma Città Futura: An Exhibition, 10 Tables, a Workshop and a White Paper

The *Urban Strategic Design (USD)* tool has been refined over time thanks to continuous verification, crossing didactic, research and dialoguing with numerous Public Administrations. This methodology finds its deepest verification in the project experimentation and sharing paths for the city of Parma. The didactic work on the city, started in 2014 with the Architectural Design Laboratories of the University of Parma and refined in the following years by a dedicated research group, produced its first important outcome in 2016 with the exhibition *Parma Città Futura.*

The exhibition of urban and environmental regeneration proposals for the city was open to the public from July to October 2016 at the former Oratory of San Quirino, in Parma. *Parma Città Futura* represents the vision of a green, accessible and welcoming city, able to recover its historical identity and reinterpret it for contemporary uses. Specifically, the project proposal redesigns the strip between the historical center and the first suburbs to create the linear park of the walls and imagine the recovery and enhancement of the historical and artistic heritage. Rationalizing the road system would open up the possibility of recovering a significant portion of public space to be reconsidered in a unitary design that would connect the city's collective venues through protected routes and crossing squares.

Alongside the presentation program of the *Urban Strategic Design* for Parma in 2016, were activated 10 thematic discussion tables. The tables addressed issues central to the future of the city: culture, sport and leisure, active citizenship, scientific communication, music, economy and territory, sustainable mobility, innovative technologies for the home and the city, health and wellbeing, and implementation strategies. The coordinators reasoned together with some figures from the University of Parma in order to understand how the different scientific disciplines could make a contribution useful to the city. External guests and actors from the territory were identified to testify to outcomes and solutions developed in similar contexts. Each table identified a series of concrete and feasible proposals, involving the city in numerous preparatory meetings for the public moment and the subsequent synthesis that gathered many contributions, directing them towards a common goal. The workshop, held between September and October of 2016, represented another fundamental moment of comparison to reflect on the city's regeneration proposals (Fig. 2).[11]

---

[11] Costi Dario, *Per un laboratorio permanente di progettazione urbana strategica*, in "Domus", *La città dell'uomo*, 1007, November 2016, pp. 16–19. Introduction by Domus Director, Nicola Di Battista: "At a time of great expectations for Italian cities, the collective work of *Parma Città Futura* proposes new design strategies capable of prefiguring the urban transformations needed, to be implemented with the close collaboration of universities, administrations and citizens".

Fig. 2 Debate between guests and students during the exhibition Parma Città Futura, 2016

The *Civil Society* mobilization process was conceived precisely as an opportunity for collective and transversal commitment: it offered Public Administrations themes to reflect upon. A key concept was accessibility, understood not only as the possibility of use spaces by all, but also as the objective of the immediacy of comprehension by the population. For this reason, a project presentation methodology was refined to be as clear as possible, capable of linking three key factors: the project's capabilities, the sharing of the proposal, and the gathering of ideas through a general mobilization.

The first aspect is the urban project, in the continuous dialectic between the identification of a system of places and the reflection on the destinations to be assumed. Secondly, the exhibition with materials designed for easy comprehension by the non-expert (photographic images comparing the actual and planned state, models at various scales and graphics with balances). The last factor was a coordinated action aimed at communicating the proposal and involving the population accompanied by an explanation of the proposed scenarios, thanks to a guided tours of the exhibition and numerous opportunities for discussion.

The citizenship involvement was developed on two levels of participation: top-down and bottom-up. The discussion of the city's issues of interest through the 10 working tables was the main instrument of a top-down approach that allowed experts to discuss key urban issues and to bring out proposals for the regeneration of public spaces. These suggestions, together with the preliminary work elaborated by the research group, were collected and presented to citizens in the exhibition itinerary. The bottom-up approach, on the other hand, was realized through a busy program of guided tours, publicity for the initiative in the local media and a photographic competition within the exhibition, entitled: *City in the middle of the city.*

Indeed, the best photos posted on Instagram with the hashtag *#parmacittàfutura* were selected in a small exposition within the same exhibition. Here the photos were divided into two sections, corresponding to the two hashtags: *#Parmachemipiace* and *#Parmachenonmipiace*, which respectively collected color photos of places with identity value and black and white photos of the city's *non-places* (Augé, 2009). The competition gave people who live Parma every day the chance to express themselves on the places they usually visit. A significant outcome of this experience is the publication of the *White Paper: Parma Future City Volume II*, which represents the collective effort to synthesize the issues and perspectives that emerged in the open sharing process, during 2016.

5.2 *Parma Città d'Oro: The Discussion with Experts and the Involvement of the Citizens*

The progress of the experimentation on Parma finds another moment of dialogue with the population in the exhibition *Parma città d'oro* held between September and December 2021 at *Palazzo Bossi Bocchi* in Parma. The initiative starts from the results achieved by *Parma Città Futura*. This project made possible to outline a strategic scenario linking the territorial scale to the urban one, the infrastructures and the landscape, the suburbs and the center in an integrated vision developed on this occasion.

The research questions the identity value of Parma in cultural, tourist and economic terms, starting from the appellation of *Crisopoli*,[12] used in the past to identify Parma as the city that produced money. This definition has, over time, taken on many declinations that grasp its urban preciousness from different points of view. Often the city and its monuments have themselves become the *golden city*, *Crisopoli* has thus become the symbol of a shared quality to be protected and of an excellence to be valorized. In order to actualize the definition of *Parma Golden City*, it was necessary to start from two connected components, the physical part and the human factor. The system of public spaces represents the contemporary *golden city,* crossing the settlement fabric and the historical-artistic heritage with a collective vocation (Fig. 3).

For this reason, the participatory experience was again based on a sharing process articulated along two converging lines. The connected actions are developed from the top and bottom: the discussion on urban identity, involving the skills and experiences present in the context and the dialogue with the population, involved in debate on possible destinations and options to be chosen.

[12] Referred to «*Iulia Crysopolis que est Parma*». Conti G, *Anonimo ravennate del VII Secolo, La città d'oro*, Libreria Ticinum, Voghera, cit. p. 3.

Fig. 3 Set up of Golden City, Exhibition Parma città d'oro, 2021

The top-down approach is developed through meetings with experts from different background and influent Parma personalities who, in the months preceding the opening of the exhibition, developed scenarios of great interest that were laid out on the city's map as integrations or alternatives. Specifically, 15 video interviews were conducted with some of the city's most influential personalities, who presented their reflections on crucial themes for Parma, such as music, literature, economics, architecture, culture, etc. Moreover, 11 working tables, held between September and November 2021, took up the results of the *Parma Città Futura* experience, updating them with new proposals for relaunching the city after pandemic.

The outcomes of this public debate became a broad reflection track, leading to refine the project proposals in the exhibition. Indeed, the video interviews were displayed alongside the new scenarios for the regeneration of the city's public spaces. These contributions were then discussed in public and offered to citizens as options to choose from.

Subsequently, a broader sharing process with a bottom-up approach was activated. In fact, in addition to project proposals for the regeneration of the city, the exhibition presents an analysis of more than 80 historic buildings of collective value. These are described in their spatial and typological quality in relation to the urban morphology and are studied to understand their potentialities, through a subdivision into three categories: *existing heritage, available heritage* and *projects.*

This tool takes the name of *Civil Atlas of Architecture* and constitutes an important innovation in the national panorama, because for the first time in Italy is carried out a census of the historical heritage with a collective vocation of a city. The *Atlas* allows public bodies, stakeholders and citizens to question themselves on the definitive destination of Parma's architectural heritage with a collective vocation.

This work is offered to the city as a proper collection of ideas and has been arranged to hear the opinions of the population. After seeing the elaborated hypotheses and hearing the words of those who have provided a contribution, the visitor reaches the heart and brain of the exhibition: the *Sharing Room.* Here, the scenarios are presented, proposing some verified and practicable alternatives, leaving the space for new ideas. In fact, visitors were able to express their preferences for the future uses of 26 buildings out of the 80 surveyed in the *Atlas*, through an anonymous form refined with the contribution of some sociologists (Fig. 4).

The goal of this moment was both to collect the choices of each visitor in order to have a reliable evaluation of the preferred uses of the citizenship and to stimulate in the participants the ability to be active subjects in the public debate about the place

Fig. 4 Sharing Room, inside the exhibition Parma città d'oro

where they live. Moreover, it was possible to visit the exhibition and take part in the consultation not only in presence but also remotely, through a dedicated website, which allowed the population to deepen their understanding of the themes dealt with in the exhibition. The participation process included also an intensive publicity campaign via social media (Instagram and Facebook), the exhibition's website and a dense program of contributions in the local media. This effort allowed to inform the public about the exhibition, in order to keep up-to-date on the various initiatives, as guided tours, public discussion panels, but also on the proposed content.

At the end of the exhibition, it was possible to understand which hypotheses were awarded and which favorite scenario emerged by the citizens of Parma. These results will be available to the Public Administration as a starting point for regeneration work aimed at restoring Parma's identity values.

6 Conclusions

The participatory experiences that supported the experimentation of the *USD* for the city of Parma, have been the demonstration that participation works when it is clearly oriented, questions people on easy-to-understand scenarios and on the possibilities of intervention.

The outcome and the numbers that marked the *Parma Città Futura* experience, which involved 62 organizations and associations and 137 people, highlighted the effectiveness of this approach. The main result, however, was the phenomenon of collective mobilization that flanked the exhibition, gathering a convergence of knowledge on the city's main topics of interest. Indeed, the exhibition *Parma città d'oro*, involved a total of 3998 visitors, among them 547 expressed their vote in presence, while only 18 people gave their preferences online. These numbers show how the social and collective component of the process is one of the elements that motivate to participate in decision. Citizen involvement is indeed effective when it takes place in presence: among the 565 voters, over 96% voted in presence.

Another significant aspect regarded guided tours and in particular school tours, which counted 1862 visitors and allowed children and young people to know the city through an itinerary designed for the youngest. The most important aspect that characterized this initiative is the cultural and scientific results achieved. This experience has provided an opportunity both to educate young people and children, thanks to a specific approach based on the immediacy of content, and to raise awareness of urban issues, involving the visitors in the choice of future destinations for the city's spaces and buildings with a collective vocation. The exhibition then described a virtuous scenario on the best European standards, through a comparison with 30 cities of similar size to Parma, that have started urban regeneration and environmental sustainability projects.

Parma città d'oro also provided the opportunity to activate a dual system of participation: from top and bottom, involved experts and citizens, engaging in the collective challenge of opening a public discussion on the future city. Finally, the

Civil Atlas of Architecture was an essential tool in the participatory process, allowing visitors to express their opinion, in the *Sharing Room*, on the possible destinations of the buildings analyzed in the *Atlas*. This is the element that determines the systematic and scientific nature of the approach, consolidating at the same time the awareness of the civic role of the visitors, who were active participants in the definition of a new scenario for public spaces with a collective vocation in Parma.

References

Arnstein, S. (1969). A ledder of citizen participation. *Journal of the American Planning Association,* 216–244.

Augé, M. (2009). *Nonluoghi. Introduzione a una antropologia della surmodernità.* Elèuthera.

Barbera, L. V. (2019). *La città radicale di Ludovico Quaroni. Ludovico Quaroni e la Scuola di architettura di Roma negli anni Sessanta.* Gangemi.

Bertelli, G. (2018). Paesaggi fragili. *Aracne editrice.*

Carta, M. (1996). *Pianificazione territoriale e urbanistia. Dalla conoscenza alla partecipazione.* Edizioni Medina.

Carta, M. (2003). *Teorie della pianificazione: questioni, paradigmi e progetto.* Palumbo.

Cassese, S. (2017). *La democrazia e i suoi limiti.* Mondadori.

Ciancaglini, M. (2010). Tra democrazia partecipativa e concertazione. La legge regionale 3/2010 dell'Emilia-Romagna. *Istituzioni del federalism, XXXII,* 215–253.

Costi, D. (2021). *Designing the city of people 4.0. Reflections on strategic and sustainable urban design after Covid-19 pandemic. For the series The City Project.* Springer Network.

Culotta, P., Leone, G., Bianchi, A. L., Sarta, G., Panzarella, M., Minutella, V., Foderà, L., et al. (1988). *Palermo 1991: Nove Approdi per l'Esposizione Nazionale.* Democrazia Cristiana.

Davidoff, P. (1965). Advocacy and pluralism in planning. *Journal of the American Institute of Planners.* The City Reader. Consulted edition Davidoff, P. (2004). Advocacy and pluralism in planning. In J. Stein (Ed.), *Classic readings in urban planning* (pp. 41–52). Planners Press.

De Carlo, G. (1971). *L'architettura della partecipazione.* Royal Australian Institute of Architects. Consulted edition Marini, S. (Ed.). (2015). *L'architettura della partecipazione.* Quodlibet.

De Carlo, G. alias Gimdalcha Ismé. (1995). Progetto Kalhesa. Consulted edition (2014). *Edizioni di Storia e Studi Sociali.*

De Jong, T. (1992). *Kleine Methodologie voor Ontwerpend Onderzoek.* BOOM Uitgeyers.

Dolci, D. (2020). *L'educazione Edizioni di Comunità.*

Franchini, A. (2020). *Il villaggio Matteotti a Terni. Giancarlo De Carlo e l'abitare collettivo.* L'Erma di Bretschneider.

Friedmann, J. (1987). Planning in the public domain: From knowledge to action. Princeton University Press. Consulted version (1993). *Pianificazione e dominio pubblico. Dalla conoscenza all'azione.* Dedalo.

Geddes, P. (1915). Cities in evolution. Williams & Norgate. Consulted version (1970). *Città in evoluzione.* Il Saggiatore.

Kotkin, J. (2020). *After coronavirus, we need to rethink densely populated cities.* Fortune, 1 April. https://fortune.com/2020/04/01/coronavirus-dense-cities-urban-rural/

Mazzoleni, D., & Sepe, M. (2005). *Rischio sismico, paesaggio, architettura: l'Irpinia, contributi per un progetto.* Doppiavoce.

McCann, E. J. (2011). Collaborative visioning or urban planning as therapy? The politics of public-private policy making. *Professional Geographer,* 207–218. https://doi.org/10.1111/0033-0124.00280

Mumford, L. (1938). *The culture of cities.* A Harvest/HBJ Book-Harcourt Brace Jovanovich Publisher. Consulted version (2007). *La cultura delle città.* Einaudi.

Olivetti, A. (1952). *Società Stato Comunità.* Edizioni di Comunità.

Pontradolfi, P. (2002). La legislazione urbanistica regionale e pianificazione partecipata. In *Cultura e prassi della partecipazione nella pianificazione delle città e del territorio, Section III by Commissione Urbanistica Partecipata e Comunicativa of Istituto Nazionale di Urbanistica, attached Dossier to n° 182 of Urbanistica Informazioni.*

Renacco, N. (1955). *Il Piano Regolatore Generale di Ivrea.* Reparto Tipografico ICO.

Robert, J. P., & Steiner, E. (1989). Styrie. La participation en question. *L'Architecture d'aujourd'hui,* 128–130.

Roggema, R. (2017). Research by design: Proposition for a methodological approach. *MDPI Urban Science.* https://doi.org/10.3390/urbansci1010002

Rossi, A. (1966). *L'architettura della città.* Marsilio.

Sharifi, A., & Amir Reza, K. G. (2020). The COVID-19 pandemic: Impacts on cities and major lesson for urban planning, design, and management. *Science of the Total Environment, 749*(142391). https://doi.org/10.1016/j.scitotenv.2020.142391

Simon, D. (2020, April 23). *Cities are at the centre of coronavirus pandemic—Understanding this can help build a sustainable, equal future.* https://theconversation.com/cities-are-at-centre-of-coronavirus-pandemic-understanding-this-can-help-build-a-sustainable-equal-future-136440

Talia, M. (2011). Il divorzio tra architettura e urbanistica: dagli anni Sessanta all'ultimo decennio. In F. Evangelisti, P. Orlandi, & M. Piccinini (Eds.), *Disegnare la città. Urbanistica e architettura in Italia nel Novecento: appunti da un ciclo di conferenze. Series Leggere e scrivere la città.* Edizioni Edisai.

Thomsen, M. R., & Tamke, M. (2009). Narratives of making: Thinking practice led research in architecture. *Communicating (by) Design,* 1–8.

Tomasetti, F. (2012). *Cambiare Rimini. De Carlo e il piano del nuovo centro (1965–1975).* Maggioli.

Chapter 3
Implementing a Participatory Design Approach to Create a Sensory-Friendly Public Space for Children with Special Needs

Lingling Deng, Surabhi Ratavjia, and Prapa Rattadilok

Abstract This chapter presented a participatory design approach for involving children with special needs in public space regeneration. Specifically, this chapter focused on children with neurodevelopmental disabilities who are hyper- or hypo-sensitive to sensory stimuli or have a unique sensory interest. The study aimed to use the participatory design framework to build a sensory-friendly square based on a case in China. The case study area is located in Ningbo city, enclosed by a rehabilitation centre catering to around 300 children with Autism Spectrum Disorders (ASD). The participatory design approach used in this study began with online questionnaires in which parents of ASD children reported information about their child's preference when visiting the square. Following the online questionnaires, living lab and focus group methods were used to involve children as key informants and design partners. The living lab equipped with virtual reality devices allowed children to experience a 'real' regenerated square and the focus group after each living lab session allowed all stakeholders to provide feedback for design iterations. Preliminary results obtained from the online questionnaires were presented, inspiring several design considerations for square regeneration. The sensory experience of children at the square would be supported via designs such as sensory stations. Common techniques for smart environments were selected to build a smarter square. This chapter not only provided regeneration recommendations for current square space in China but also introduced a feasible framework for actively involving children with special needs in the regeneration process.

L. Deng (✉)
School of Computer Science, University of Nottingham Ningbo, Ningbo, China
e-mail: lingling.deng2@nottingham.edu.cn

S. Ratavjia · P. Rattadilok
School of Computer Science, University of Hertfordshire, Hertfordshire, UK

E. Mangi et al. (eds.), *Innovative Public Participation Practices for Sustainable Urban Regeneration*, Urban Sustainability, https://doi.org/10.1007/978-981-99-9595-0_3

1 Introduction

Entering a public space for people who have special needs could be particularly challenging. For example, people with anxiety and depression sometimes experience distress in an unfriendly environment. Fear of judgement and stigma towards melt-down makes them opt to hide from the public. Better healthcare services in metropolitan areas intensify the rural-to-urban migration of people with special needs. Designing and developing an inclusive and friendly public space for them is an important target for contemporary urban regeneration. Architectural guidelines and standards have been developed to take into account the needs of individuals with varying degrees of disabilities, ensuring that these individuals can access indoor or outdoor spaces freely and safely (Nagib & Williams, 2017). However, unlike indoor spaces where healthcare services and interior design can be improved and standardised, outdoor space regenerations are usually more complex, affected by environmental factors, such as background clutter, lighting conditions, facility (technology) selection, and placement.

It is noticeable that a number of children with neurodevelopmental disabilities, such as Autism Spectrum Disorder (ASD), Attention Deficit/Hyperactivity Disorder (ADHD), and other disorders, respond to sensory information differently from peers that are typically developed in daily life. They may be hyper-sensitive or hyposensitive to external sensory stimuli, such as light, touch, sound, and many more, which is described as atypical sensory processing according to the Diagnostic and Statistical Manual of Mental Disorders (American Psychiatric Association, 2013). The distress caused by sensory stimuli is likely to result in self-injurious and aggressive behaviours (Javed et al., 2019). These challenges can, in turn, lead to the social isolation of the child in a public space (Schaaf et al., 2011). It is beneficial if the public environment around these children can be sensory-friendly and equipped with some technological support. At the same time, people with special needs usually benefit the most from a participatory design process (Frauenberger et al., 2011), which involves them and has their voice to make meaningful contributions to public space modifications. Therefore, this chapter aims to describe the application of a participatory approach to regenerate a suitable public space for children with special sensory needs.

2 Background Knowledge

To understand the potential gaps for public space regeneration, we provide background with available design guidelines and current situations of outdoor spaces, such as squares and streets in the urban areas of China and the UK. Tola et al. (2021) mentioned that the surrounding environment can be problematic for people with neurodevelopmental disabilities. According to the study, people with ASD found it difficult to adapt to the activities which were planned. There was an increase in the awareness and importance of reconsidering the surroundings and urban design fields. This research suggested that it is important to re-organise the built environment which impacts people with ASD. For autism-friendly spatial requirements, colour coding and labels on the paths can support people with ASD. There should be sensory gardens for ASD children and adults in the outdoor environment and these surroundings should have some consistent elements in order to build a predictable pattern.

When designing for children with atypical sensory processing, many factors related to sensory experience should be considered. Firstly, the colour and material used in designing should have the right balance and there should not be bright, shiny surfaces. The design should also avoid busy patterns and heavy textures. In addition to the balance of colour and material, the lighting approach should be layered which can help limit the glare and reflectivity within the space resulting in providing visual comfort. Furthermore, the design should consider introducing acceptable sound levels in the area. Considering sensory zoning within the area which can divide the children into different stimulus zones (Walczak, 2020). Every space, area and design should be safe as children with ASD may have self-harming tendencies.

There were a range of non-standardised design guidelines for ASD people available in the architecture field. Knowlton School of Architecture (2018) introduced an ASD planning and design guideline for the infrastructure in public spaces such as downtown, urban, suburban, retail space, campuses, parking access, and multi-model hub. One of the suggestions was to keep the bike lanes painted green using soft glow-in-the-dark paint to increase visibility in the downtown, retail and suburban areas. It was suggested that crowd on the urban sidewalks can cause anxiety and hence, the walking section should have a magenta thermoplastic strip at the centre. Knowlton School of Architecture (2018) guided to keep the lanes for travel narrow, which can help people with autism less anxious. This guideline mentioned keeping the drive lanes 10 feet wide and the streets to be multi-modal. The announcements for the bus route can cause stress and anxiety for adults with autism. During the travel journey on the bus, there should be a visual or audio message so that the passengers were aware of the stop. It also suggested having the 'I Need an Assistance Symbol' on the inside and the outside of the bus complying with the checklist guideline. According to a guideline developed by Pedder and Scampton Architects (2017), the education buildings for ASD children should identify the levels of need of groups so that a particular level of support can be provided. It suggested avoiding roadways with heavy traffic, railways, airports or any other commercial places. According to Gaudion and McGinley (2012), the external noises from the street can be avoided by designing natural sounds such as sounds of water, sounds of bird songs, and many more.

Figure 1 is a public square located in front of a rehabilitation centre in Ningbo, a 'median economic level (neither extremely affluent nor extremely poor)' city in southeast China (Sun et al., 2019, p. 3). The square is equipped with playground areas, plants, parking slots, and indoor rehabilitation space, near a six-lane major roadway with convenient transportation. As a square close to the rehabilitation centre,

Fig. 1 Satellite aerial image (left) and street view (right) of a public square near a rehabilitation centre for children with special needs in Ningbo, China (*Source* Apple Maps (left), Baidu Maps (right))

it has been frequently visited by children and adults with ASD, ADHD, and other disabilities. Although the design of the square has considered the sensory needs of children by using coloured facilities and natural light conditions, there are several obvious weaknesses. For example, the walking path of the square has always been crowded because the space for walking is squeezed by side parking slots. The square is located near a very busy road but lacks enough protection to prevent unpleasant noise.

Figure 2 shows a space near a school for children with special needs in the UK which is accredited with Autism Accreditation. This school has a soft play which is a recreational activity area inside the building and has nine different classroom areas. It has an outdoor play area having a playground covered with a small circular road for driving secured by a fence. The nearest circle to this school can be seen in Fig. 2 (right). From the sensory design perspective, it is visible that the circle has been built in a standard way however, there is a scope for improvement by considering the necessities of children with special needs such as, colour painted stripes on the road, light conditions during the day and the night as some colours in the day or night can make children with special needs anxious and, a safe sidewalk with an emergency alarm system which should be specifically built for the children with ADHD to operate. Similarly for this site, the road is in front of the school building which may or may not be busy at times however, this can affect the children and hence, there should be a system installed to control the noise and the crowd near

Fig. 2 Satellite aerial image of a public space near a school for children with special needs (left) and street view of the nearest circle and road to the school (right) in the UK (*Source* Google Earth (left), Google Maps (right))

the school area. The school management and the nearby areas should consider re-building and re-designing the road, circle, and school premises in order to fulfil the sensory needs of children.

Figure 3 displays a busy crossroads in London. This area has been well built and it suffices a few of the sensory requirements for children or adults with ADHD and ASD such as it has colour painted stripes on the road and has enough wide space of the footpath which can be seen in Fig. 3 (top). Sound and noise can cause anxiety in children and adults with ADHD and ASD and this sensory factor should be re-considered. There should be separate rooms built for children and adults with ADHD and ASD besides such busy and crowded roads so that children and adults can calm down and rest from sudden anxiety.

3 Participatory Design and Technology for Space Regeneration

Participatory design is subtly different from 'participation', the concept in which stakeholders are usually involved as users and testers (Halskov & Hansen, 2015). In many previous studies, one of the most common types of participation is the involvement as users and testers, with the goal of evaluating the design. Further to these studies, a more inclusive model of participation, defined as participatory design, suggests the appropriate involvement of stakeholders as informants or design partners in the design process to help ensure that their needs are met (Dell'Era & Landoni, 2014; Druin, 2002; Guha et al., 2008). The role of users and testers is usually passive compared to the role of informants and design partners which carry the notion of active participation (Druin, 2002). As informants, individuals with special needs can

Fig. 3 Street view (top) and satellite aerial image (bottom) of a busy crossroads in London (*Source* Google Maps (top), Google Earth (bottom))

be involved at certain stages for particular reasons, such as in the process of low-tech concept design and feedback on particular design decisions. As design partners, they are recognised as equal stakeholders in the design process and involved throughout the process.

Recently, participatory design has been used in many new domains due to technological and societal development. Urban regeneration is one of the new domains where this approach is employed incorporating multidisciplinary contributions from architecture, engineering, informatics, sociology, psychology, and other research fields. The rural-to-urban migration of families of children with special needs requires the corresponding urban designers to perceive them not only as vulnerable but also as citizens and partners. Knowledge about their needs for space regeneration is traditionally acquired from questionnaires and focus group interviews (Małgorzata, 2020). For example, when designing public spaces for children with special sensory needs, sensory-based questionnaires were used to understand their sensory limitations and preferences. Some studies investigated the impact of various environmental factors, such as noise, light, temperature, and humidity, by involving children with neurodevelopmental disabilities, and thereafter, proposed designs such as learning environments (Khare & Mullick, 2009; Mostafa, 2008; Scott, 2009) and residential-educational centres (Beaver, 2010).

The living lab method is emerging, which provides new perspectives for participatory design. It can be defined as 'a real-life testing and experimentation environment' where users are aware that they are involved in the co-design process (Dell'Era & Landoni, 2014, p. 4). A living lab contains one key element which is the simulation of real-life settings. Although living labs rely largely on indoor facilities with which it is hard to simulate public environments, the development of virtual reality (VR) makes it possible to implement a safe and 'real' public space design in a lab setting. Programmable VR devices on the market, such as Oculus Rift, allow participants to undergo different virtual experiences according to their preferences. Such devices have been used in recent studies which showed the feasibility and acceptance among children with neurodevelopmental disabilities in living use (Farroni et al., 2022; Johnston et al., 2020).

It has shown that technological innovation facilitated a greater degree of collaboration between designers and technological manufacturers who used to only serve as skill and material providers. A typical example is the ECHOES project led by researchers who came from a wide range of research departments. The project implemented a participatory design that involved children with special needs in the design process, aiming to create a technologically enhanced learning space (Frauenberger et al., 2011). They designed a software system with an interactive interface to explore the sensory preferences of children and to provoke imaginary play. Some engineering research worked with no participatory practice but provided novel technological solutions for space management. For example, smart lighting systems equipped with sensors and actuators have been widely tested and discussed in urban planning to ensure that the system provides safety and comfort to citizens (Mahoor et al., 2020).

4 Methodology

In this chapter, a participatory design framework is proposed to regenerate a sensory-friendly public space, primarily a square for children with special needs. The tools used in this context include online questionnaires to involve participants regardless of geographical location, a living lab method to facilitate meaningful participation, and focus group interviews to obtain feedback.

4.1 The Case Study Area

The area taken as a case for regeneration is the square in Ningbo as shown in Fig. 1, which is partly developed. The square is small in dimension compared to the large public space, which is about 5.5 acres. There used to be a fountain in the centre of the square. It has been demolished to provide more space for the new design. The area plays a crucial role because it is enclosed by a rehabilitation centre for people with neurodevelopmental disabilities. As introduced by the coordinator of the rehabilitation centre, there were about 300 children with ASD, mainly young-aged. Therefore, the regeneration of the area should strengthen a better relationship between the square and its nearby ASD community, providing a sensory-friendly space for children with ASD.

4.2 Online Questionnaire

An online questionnaire was designed to obtain preliminary information about the needs of ASD children at the public square, including questions about children's age, frequency of square visits, and preference for square features (see Table 1). The online questionnaire was conducted anonymously and was distributed through the rehabilitation centre in the case study area. Parents of children diagnosed with ASD under the age of 18 were sampled. A Chinese version of the questionnaire in Table 1 was used. This procedure was reviewed and approved by the Research Ethics Committee of the University of Nottingham Ningbo China. The results of the online questionnaires helped the authors to determine the sensory stimuli and features to be used to design a desirable square for children. Although questionnaires are traditionally used in the initial phases of design to identify needs, it cannot completely probe into a subject's needs as he or she may not know the needs that he or she may manifest in the further phases. Therefore, additional tools including a living lab and focus groups were also employed.

4.3 Living Lab

A living lab will be kitted out inside the rehabilitation centre, equipped with VR devices, cameras, and a touch station. VR devices will be used to realistically simulate a 3-dimensional public square. VR developer platforms such as Unity can be used to program the scene with images of the real world, paired with sound effects which simulate the traffic noise. Preferred features reported by parents of ASD children will be considered in the regeneration simulation. However, to know the true needs of children, object properties such as color, shape, texture, and sound will be adjusted by several editors in the VR program. A physical touch station, including a demo

Table 1 Questionnaire for sensory-friendly public space regeneration

Question description	Options/sub-questions
1. What age range does your child fall between?	0–2; 3–5; 6–12; 13–18
2. How often do you and your child visit your local square?	Daily; weekly; monthly; rarely
3. How would you normally get to the square?	Walk; bike or e-bike; car; bus; underground
4. What does your child like to do at the square? (Check all that apply)	Play on the playground equipment; play on the swings; go for a walk; sit and read; enjoy nature; play sports (basketball, soccer, baseball, etc.); go skateboarding; other
5. What do you look for at a square to make your child's visit more enjoyable or comfortable? (Please choose THREE of the most important things to you)	Indoor space; paths—opportunities to walk and exercise; seating; variety of playground features that appeal to children; music; artificial lights; water features such as fountains; textured walls; paintings; trees and plants
6. What is your child's preferred combination of colour for public space facilities? (Please choose THREE of the most favourite colour)	White; lilac; violet; dark blue; blue; light blue; turquoise; green; light green; yellow; orange; pink; red; dark red; brown; grey; black; no preference; other
7. What is your child's preferred sound in the public space? (Check all that apply)	Soft music; pop music; nursery rhyme; sound of blowing wind; sound of rain drops failing; sound of thunder; sound of a waterfall; sound of traffic; sound of chirp; no preference; other
8. What is your child's preferred texture in the public space? (Check all that apply)	Smooth texture; soft texture; ice cold texture; plastered texture; stone texture; rubbery texture; leaf texture; no preference; other
9. What is your child's preferred shape in the public space? (Check all that apply)	Circle; rectangle; triangle; square; oval; hexagon; cylinder; sphere; cube; no preference; other
10. From 1 (not at all) to 5 (very much), at a square, how much does the following circumstance make your child feel uncomfortable? (Provide rating based on 5-point Likert scale)	Too much light
	Very low light
	A lot of people
	A lot of noise
	Strong smells
	Narrow, small space
	Large space

textured wall using different sets of materials, a soft blanket, and a bowl of sand, will be put in front of the participant to simulate potential tactile experience at the square. Cameras will be installed to capture the participants' interaction with the design.

Participants for the living labs will be pre-selected and invited based on the inclusion criteria. One of the inclusion criteria is that the children should be formally diagnosed with ASD or other neurodevelopmental disabilities. The inclusion criteria should also consider the age of participants. Only children aged 13 and above will be invited as the VR headset may cause discomfort for small children due to improper sizing (Oculus, n.d.). Besides, the parents of the child should give consent for the involvement and monitor children who are participating in the testing to ensure their safety. The living lab allows participants to understand the impact of regeneration and their true needs after experiencing the design. Through a living lab, designers can observe the practicality of the design and participant acceptance in a simulated environment.

4.4 Focus Group Interviews

There will be a focus group after each living lab session to allow designers, children participants, and their parents to explore the impacts of the regeneration and the use of sensory stimuli. Children will be required to apply an odd-one-out strategy to rule out the features that they dislike. Both negative and positive feedback will be used to influence the future design of the public space which needs regeneration. The living lab scenes will be adjusted iteratively to meet the needs of children.

5 Preliminary Results and Design Considerations

5.1 Questionnaire Results

The online questionnaire finally received a total of 91 responses. Most of the parents reported that their children visited the square weekly (61.5%). Very few parents (2.2%) indicated that they rarely went to the square. According to the questionnaire results, these parents carried the children to the square mostly by bike or e-bike (37.4%), or by car (33.0%). The most favourite activities of ASD children at the square were playing on the playground (69.2%), playing on the swings (68.1%), and going for a walk (36.3%). Therefore, adequate playground features were mostly concerned by the parents (87.9%). Water features (51.6%), music (30.8%), and textured walls (30.8%) at the square can also make children comfortable and enjoyable. In terms of visual elements, ASD children preferred blue (44.0%), green (41.8%), and red (33.0%). For auditory stimuli, nursery rhyme (73.6%), soft music (34.1%), and pop music (26.4%) yielded the highest frequency among all the options. Many children preferred soft texture (38.5%), smooth texture (31.9%), or did not have any specific preference (27.5%). Circles (45.0%), spheres (19.8%), and rectangles (19.8%) were preferred in terms of selecting the shape of an object. Regarding the level of discomfort, parents reported that 'too much noise' (mean score: 3.08), 'strong smell' (mean score: 3.07), and 'narrow, small space' (mean score: 3) bothered their children more significantly, obtaining highest averaged rating on Likert scale.

5.2 Design Considerations

When designing the square for the children, several factors were taken under consideration pertaining to the sensory preferences of children, the current conditions of the square, and the state-of-the-art technologies. Table 2 summarised the design considerations.

The design involves several sensory considerations, intending to provide positive experience in visual, auditory, tactile, olfactory, and spatial senses. Most of these considerations can be firstly addressed within the VR environment. The touch station in the living lab can simulate the tactile experience as described in Table 2. In real development, different sensory stations should be arranged at different locations within the square so that children will not be given too many stimuli at the same time.

A quick response (QR) code will be provided at each entrance to the square, linking to a mini program that provides introductions of available facilities, sensory stations, and parking information about the square. The mini program will also be a platform for integrating ambient sensors, providing a dashboard (such as Fig. 4) for parents to know the brightness level, noise level, and temperature/humidity level. Besides, the information collected by the sensors is the key to knowing the routines or ritualised patterns of movement of the children within the square (Polo-Rodríguez et al., 2021). By analysing the presence data, researchers and designers will be able to know which stations are more frequently visited in real life and whether it is necessary to increase the number of stations to avoid crowding.

6 Discussion and Implications

In this chapter, the authors proposed a participatory design framework for public space regeneration for children with special needs. Some initial explorations were carried out. However, the current study has not moved on to the implementation of the design. Through the review of background knowledge and actual engagement with ASD children, authors identified that ASD children had specific sensory preferences in the public space and their sensory needs were not fully met in the current context. The preliminary findings assisted the authors to design new features for a public square regeneration project in China.

On the other hand, this chapter reflects the significance of the participation of children with special needs in urban regeneration. In particular, for children with neurodevelopmental disabilities, their difficulties in social situations and communication were so profound that their needs may not be completely understood. Parent-reported needs or preferences of children via questionnaires and interviews have limitations in real-life practice. Previous research suggested that it is children, rather than adults, who should become key informants in the design process (Druin, 2002). By using the participatory design framework proposed in this chapter, children can

Table 2 Design considerations for public square regeneration

Design consideration	Description
Visual experience	The square should be designed with painted footpaths in their preferred colours, like blue and green. A textured wall should be built using art glass of warm colours (e.g., yellow and red) infusing the visual zone with warmth and energy. The painted footpath and textured wall provide visual stimuli at the square. A soft, eye-protecting autonomous lighting system will be deployed for brightness supplies when the brightness level is low at the square
Auditory experience	The square must incorporate an auditory-friendly design in regeneration since 'too much noise' is one of the most uncomfortable factors that affected children. More trees should be planted to serve as barriers that reduce noise from the busy roads. Several music stations will be arranged within the square. By pushing the button on a music station, nursery rhymes, soft music or pop music will be played via a speaker to reduce the anxiety of children
Tactile experience	A glass-made textured wall will be built to provide both visual and tactile stimuli. Besides, several touch stations equipped with soft furry blankets and sandbox will be placed within the square. The materials used for tactile experience are those which are more sustainable in open spaces
Olfactory experience	Preliminary evidence shows children's aversion to strong smells. The floor of the square will not use materials that may smell unpleasantly such as plastic grass. Tiles and cement will be used instead. A flower bed will be developed with real flowers and scented artificial plants to provide olfactory stimulus. The scent could be modified based on the season. For example, the scent of sweet osmanthus could be provided during the autumn
Footpath regeneration	The square needs wider pedestrian footpaths. The current parking slots by the side of footpaths will be removed to meet the spatial needs of children with ASD. There will be an independent car park and parking space for bikes (as well as e-bikes) adjacent to the square so that the zone within the square will be car-free to ensure the safety of the children. Pedestrian footpaths will be coloured
Choice of techniques for a smart square	Information and communication technology (ICT) and common sensors for smart environments will be used, including: • Presence detection. Sensors of presence will be placed in some target locations such as footpaths and sensory stations to detect human presence. The presence of children at a certain location can trigger stimuli automatically. For example, children showing up at the flower bed can trigger olfactory stimuli • Ambient sensors such as light sensors, light control, temperature/ humidity sensors, and sound sensors. Ambient sensors can provide information about the environment to children and their parents, avoiding a potentially uncomfortable outdoor situation

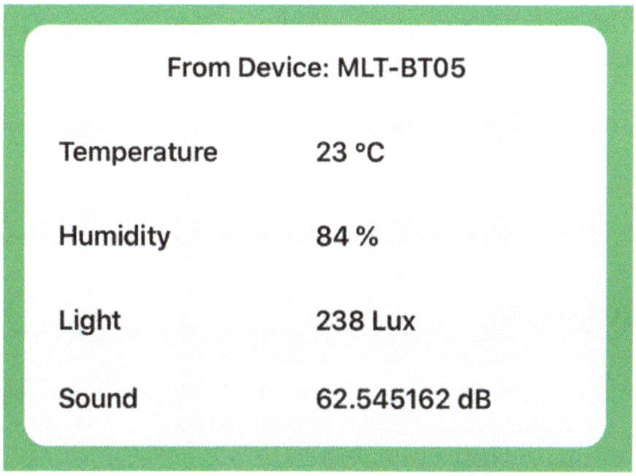

Fig. 4 A demo ambient sensor data dashboard developed by the authors

be actively involved, beginning to show their true needs for the preferred features that correspond closely to living experiences.

Although traditional methods such as questionnaires and interviews can be successfully undertaken with other stakeholders (e.g., parents and teachers) during the earliest stage of the design, these 'low-tech' development activities were believed to be unrealistic for ASD children (Keay-Bright, 2007, p. 8). Since the importance of technology in participatory design has been increasingly recognised, this chapter incorporated an emerging method, the living lab, in the framework using smart techniques. The application of VR in the living lab provides an innovative implication for future participatory design research. Future participatory design practice can also adopt other ICT techniques and hardware devices in the living lab. For example, smart wristbands are a kind of non-invasive sensor that can monitor a user's physiological data, such as heart rate and skin temperature, to identify nervousness and anxiety. The use of smart devices and analysis algorithms may identify more accurately whether the design causes discomfort or anxiety.

7 Conclusions

The participatory design framework proposed in this chapter has provided a methodology for enrolling an ASD community in urban space regeneration. Children and parents were actively involved at different stages of the design. Knowledge about children's sensory preferences and needs was obtained from parents based on their daily observations. The framework suggested that children should take an active role in the design process as key informants and co-designers. Therefore, a living lab will be helpful as it provides a safe place for children to experience the regenerated square within a VR-simulated environment. Although the initial findings have shaped the design, further efforts are needed to implement the living lab and focus group interviews. Prototyping of the technological system for the smart square will be developed.

These efforts will finally result in a sensory-friendly, usable, and sustainable public space that meets the demands of children with special needs.

Acknowledgements We would like to thank Elim Autism for their ongoing support and collaboration.

References

American Psychiatric Association. (2013). *Diagnostic and statistical manual of mental disorders* (5th ed.). American Psychiatric Publishing.

Beaver, C. (2010). Autism-friendly environments. *Autism File, 34*, 82.

Dell'Era, C., & Landoni, P. (2014). Living lab: A methodology between user-centred design and participatory design. *Creativity and Innovation Management, 23*(2), 137–154.

Druin, A. (2002). The role of children in the design of new technology. *Behaviour & Information Technology, 21*, 1–25.

Farroni, T., Valori, I., & Carnevali, L. (2022). Multimedia interventions for neurodiversity: Leveraging insights from developmental cognitive neuroscience to build an innovative practice. *Brain Sciences, 12*, 147.

Frauenberger, C., Good, J., & Keay-Bright, W. (2011). Designing technology for children with special needs: Bridging perspectives through participatory design. *CoDesign, 7*(1), 1–28.

Gaudion, K., & McGinley, C. (2012). *Green spaces outdoor environments for adults with Autism.* https://www.kingwood.org.uk/wp-content/uploads/2020/08/green_spaces.pdf

Guha, M. L., Druin, A., & Fails, J. A. (2008). Designing with and for children with special needs: An inclusionary model [Conference presentation]. In *The 7th International Conference on Interaction Design and Children*, Chicago, IL, United States.

Halskov, K., & Hansen, N. B. (2015). The diversity of participatory design research practice at PDC 2002–2012. *International Journal of Human-Computer Studies, 74*, 81–92.

Javed, H., Burns, R., Jeon, M., Howard, A. M., & Park, C. H. (2019). A robotic framework to facilitate sensory experiences for children with Autism Spectrum Disorder: A preliminary study. *ACM Transactions on Human-Robot Interaction, 9*(1), 3.

Johnston, D., Egermann, H., & Kearney, G. (2020). Sound fields: A virtual reality game designed to address auditory hypersensitivity in individuals with Autism Spectrum Disorder. *Applied Sciences, 10*(9), 2996.

Keay-Bright, W. (2007). The reactive colours project: Demonstrating participatory and collaborative design methods for the creation of software for autistic children. *Design Principles and Practices: An International Journal, 1*(2), 7–15.

Khare, R., & Mullick, A. (2009). Designing inclusive educational spaces with reference to autism. *Proceedings of the Human Factors and Ergonomics Society Annual Meeting, 53*(8), 517–520.

Knowlton School of Architecture. (2018). *Autism planning and design guidelines 1.0.* https://www.housingonline.com/wp-content/uploads/2019/11/Autism-Planning-Design-Guidelines-December-2018.pdf

Mahoor, M., Hosseini, Z. S., Khodaei, A., Paaso, A., & Kushner, D. (2020). State-of-the-art in smart streetlight systems: A review. *IET Smart Cities, 2*(1), 24–33.

Małgorzata, B.-Z. (2020). Architectural design and methods inspired by children's culture and creativity [global culture and spatial individualization]. *IOP Conference Series: Materials Science and Engineering, 960*(2), 022097.

Mostafa, M. (2008). An architecture for autism: Concepts of design intervention for the autistic user. *International Journal of Architectural Research: Archnet-IJAR, 2*, 189–211.

Nagib, W., & Williams, A. (2017). Toward an autism-friendly home environment. *Housing Studies, 32*(2), 140–167.

Oculus. (n.d.). *Product description.* https://securecdn.oculus.com/sr/oculusrift-warning-english

Pedder and Scampton Architects. (2017). *Design and autism.* https://thespaces.org.uk/wp-content/uploads/2017/07/SD17-WS-Designing-for-autism.pdf

Polo-Rodríguez, A., Rotbei, S., Amador, S., Baños, O., Gil, D., & Medina, J. (2021). Smart architectures for evaluating the autonomy and behaviors of people with autism spectrum disorder in smart homes. In A. S. El-Baz & J. S. Suri (Eds.), *Neural engineering techniques for Autism Spectrum Disorder* (pp. 55–76). Academic Press.

Schaaf, R. C., Toth-Cohen, S., Johnson, S. L., Outten, G., & Benevides, T. W. (2011). The everyday routines of families of children with autism: Examining the impact of sensory processing difficulties on the family. *Autism, 15*(3), 373–389.

Scott, I. (2009). Designing learning spaces for children on the autism spectrum. *Good Autism Practice (GAP), 10*(1), 36–51.

Sun, X., Allison, C., Wei, L., Matthews, F. E., Auyeung, B., Wu, Y. Y., Griffiths, S., Zhang, J., Baron-Cohen, S., & Brayne, C. (2019). Autism prevalence in China is comparable to Western prevalence. *Molecular Autism, 10*, 7.

Tola, G., Talu, V., Congiu, T., Bain, P. A., & Lindert, J. (2021). Built environment design and people with Autism Spectrum Disorder (ASD): A scoping review. *International Journal of Environmental Research and Public Health, 18*(6), 3203.

Walczak, M. (2020). *Five important factors when designing for autism spectrum disorder in a pediatric environment.* https://www.healthcarefacilitiestoday.com/posts/Five-important-factors-when-designing-for-autism-spectrum-disorder-in-a-pediatric-environment--24262

Part II
Participation in Built Heritage Regeneration

Chapter 4
The Regeneration of the Urban Commons: An Opportunity for Circular Cities

Luca Lanzoni and María Jesús Peñalver Martínez

Abstract In most contemporary cities there are often buildings and open spaces, both public and private, which are dismissed, empty, not used, or abandoned for several reasons. These places might be in several different sectors of the city like historical centre, suburban areas, deprived neighborhoods, etc. The rehabilitation and the management of these underused and/or dismissed places (urban commons), through collaborative processes, is a recognised way to catalyse regeneration offering social and cultural services with positive side effects in terms of jobs creation, social inclusion, socio-cultural growth, strengthen of local identities, appropriate management, and maintenance of such places, avoiding urban sprawl. In this framework, public leverage is essential to promote such collaborative processes, and for this reason, it is important to identify and understand the method to follow and find the appropriate and necessary resources. Despite their relevance, these kinds of practices are still experimental, and there is no systematic application because even if local authorities recognise the added value of such collaborative processes, promoting such practices is not easy for many reasons, mainly:

- Unclear local urban regulation (i.e., legal framework,) to promote the cooperative management of the urban assets (sleeping giants).
- Insufficient human resources and inadequate skills to manage co-creation activities.
- Lack of a design approach to urban regeneration focused on social innovation.

To be able to manage this complexity we must develop regeneration actions, not only through collaborative approaches from the bottom up, capable of making different urban and territorial actors interact but above all also with horizontal actions, capable of making different departments of local authorities in charge for the urban management cooperate, therefore design integrated urban regeneration processes.

L. Lanzoni (✉) · M. J. P. Martínez
Department of Architecture and Building Technology, Technical University of Cartagena, Cartagena, Spain
e-mail: luca.lanzoni@edu.upct.es

E. Mangi et al. (eds.), *Innovative Public Participation Practices for Sustainable Urban Regeneration*, Urban Sustainability, https://doi.org/10.1007/978-981-99-9595-0_4

1 From Linear City to Circular City

To be fully understood, the policies devoted to the regeneration of the urban assets must be inserted within the definition of the Circular City, a definition that has only recently become globally established, emerging from the broader and more treated field of research and applications referred to circularity in the urban field (Marin & De Meulder, 2018). As the most important think tank engaged on this issue, the Ellen MacArthur Foundation (EMF), points out, "*the exploration of the circular city has just begun. Important questions still need to be addressed, and a structured evidence base needs to be developed in order to enable and encourage urban decision makers to drive the transition forward in the right direction*" (Ellen MacArthur Foundation, 2017). At the moment it is not even possible to affirm the existence of a universally accepted and shared definition of Circular Economy, in which the definition of the Circular City must be inserted, even if at least for the European Union we can refer to the contents of the Action Plan proposed by the Commission in December 2015 (European Commission—COM (2015) 614/2, 2015), and then recalling the definition provided by the EMF which proposes the schematization of the so-called ReSOLVE model: Regenerate, Share, Optimize, Loop, Virtualize, Exchange (Ellen MacArthur Foundation, 2015). To better understand the ReSOLVE model below are the concepts.

- **Regenerate**. The concept of regeneration is close to that of renewable energy. This action suggests the use of renewable sources for energy and materials used as inputs while also designing a reintroduction to the ecosystem (e.g., bio-waste reinserted in the biosphere).
- **Share**. Sharing starts with simple actions like reuse or "second-hand" shopping. However, new business models create value by providing products (accurately designed to last longer) which can be used by several users.
- **Optimise**. Continuous improvement must be pursued within a circularity concept, for example by increasing the performance of products/services with the help of new technologies and an ad hoc design that aims at reducing waste in all the lifecycle phases.
- **Loop**. This action fully incorporates the concept of circularity, which is maintaining resources in the life cycle for as long as possible. Remanufacturing, recycling, anaerobic digestion are only a few examples of the most used circular strategies.
- **Virtualise**. Virtualization exploits emerging technologies in the IT field for transforming physical products in digital ones (e.g., books, music, sopping and many other).
- **Exchange**. Using new materials, technologies, products/services for exchanging old practices with new and more circular modes can enable a disruptive transition towards more sustainable and long-lasting production and consumption.

Obviously, this schematization of circularity comes from studies of ecology and environmental economics, and how to affirm Prendeville et al. (2018) badly fits the principles and models of circularity in the field of urban sustainability. This aspect represents an important challenge because we are faced with a new and further adjective of urban sustainability policies that are added to many other definitions that emerged and have been implemented in three decades. The article by Prendeville, Cherim and Bocken is also useful for the chronological reconstruction of "urban environmentalism", for the rich bibliography relating to the main international initiatives, summits and networks of cities active in various policies, such as: the World Mayors Councilon Climate Change (WMCCC); the Cities for Climate Protection Network; cities engaged in the promotion of the so-called "urban metabolism" and "zero waste" or for the creation of smart cities; the definitions of sustainable cities, green cities and eco-cities. The article proposes important criticisms of the "Circular Economy approach for cities", for the "over-emphasis" dedicated to the recycling of materials, energy and resource flows and the excess of attention dedicated to the issue of efficiency and competitiveness, rather than that of environmental preservation, impoverishing the political debate about what sustainability really means in the city.

In fact, in the face of a quarter of a century of good sustainable development practices (Agenda 21, transport, waste treatment, etc.) widespread in most cities of the world, despite the production and implementation of policies, programs, plans and projects for the smart city (energy, cabling, digitalization of public administrations, etc.) or for the green city (urban forests, urban ecological networks, urban water management systems, etc.), despite the enormous amount of writings, scientific, programmatic or popular on these issues, the results in terms of sustainability are very poor if it is true that we had to give ourselves two deadlines—2030 and 2050—in order not to escape from the so-called Planetary Boundaries. But all these "urban sustainability" actions, aimed at specific sectors, have in any case excluded the issue of the adaptive reuse of the urban fabric and its main components: buildings and open spaces, which represent the beating heart of our cities.

It is not just a matter of thinking about technical solutions, which can make the urban fabric less impactful, but also of understanding how to involve the inhabitants in identifying how to reuse parts of the city that are not used or have been forgotten and put them back into operation in a consistent with the needs of the city. It is therefore clear that in addition to the issues of reuse and recycling of the city's products, we must also start a serious reflection on the mechanisms of inclusion of citizens with respect to the reuse, revitalization, and regeneration of the urban commons. The urban essential facilities, when governed collaboratively by different urban actors, can be conceptualized as "urban commons" or "neighborhood commons". The urban commons are tangible and intangible assets, services, and infrastructures that are collectively managed, governed, and eventually owned.

When a co-governance approach promotes the alliance between relevant urban actors and creates an ecosystem in which neighborhood or urban essential facilities are the starting point for a process or urban sustainable development, for a just and democratic city strategy, we can talk about a "city as a commons", or "co-city" (Foster & Iaione, 2016).

1.1 The City Has Always Been Circular

If we look more closely at the history and evolution of the city, it makes sense to use the concept of circularity in a natural way. The city throughout history, in fact, has always been like an "organism", a system of material type (buildings, open spaces, streets network, etc.) and flows (citizens, users, commercial functions, work activities, etc.), a complex human artifact that has simultaneously followed processes of linearity (demographic, economic, spatial growth) and circularity (adaptation, reuse, transformation of the existing). Today the city still follows this double trend, with a continuous growth on the spatial level (and a continuous consumption of the soil resource) and an incessant process of internal redefinition, which concerns urban areas, large building complexes, single buildings, commercial and production areas. Since the application to the city of the concept of transition to circularity is still recent, its application to urban cases of the past could result in an anti-historical conceptual operation; however, it can be considered a legitimate operation as the city is intrinsically circular. Just to give an example, think of how historic centres have been reused, especially the Italian ones since the 1960s and 1970s to promote social housing. The city of the past, for which we cannot speak of urban planning, can be considered an example of urban reuse and adaptation.

Faced with the adaptability of the urban fabric of historic centres it seems that the contemporary city, built from the beginning of the last century and then, with great impetus, from the mid-1950s onwards, has lost the ability to recycle, reuse and re-adapt, in particular the more recently manufactured objects: large public residential complexes but, above all, large areas and industrial buildings or large infrastructural nodes now obsolete.

Faced with the need for functional modernization, the contemporary city generally follows the path of replacement and delocalization through the abandonment of the existing one, opting for new construction, and continuous consumption of land on the edge of the compact city. Recycling, reuse, and urban adaptation in these cases take place over an excessively long period (for Italy, think of the industrial areas that were abandoned during the 1980s and only partially recovered. For Spain, think of all the real estate assets built but never used since the 2008 crisis).

1.2 The Long Journey of the Term Re-

Starting from the early 1970s, the disciplines of architecture and urban planning had begun to define new theories and practices characterized by the prefix *Re-*, highlighting the need to improve these actions: restoration, recovery, restructuring, reuse and then redevelopment, revitalization, rehabilitation, regeneration are terms, especially the last four, which established themselves since the 1980s in all countries of the Western world.

Recently the PBL—Netherlands Environmental Assessment Agency defined a real *R-strategy*, broader than the typical R-practices of architecture and urban planning because it is open to the world of objects and industrial production. The terms proposed by the latter are: Refuse, Rethink, Reduce, Reuse, Repair, Refurbish, Remanufacture, Repurpose, Recycling, energy Recovery (Potting et al., 2017). This process of circularity presents similarities between what is spreading in industrial production now and that which is partly traditional, and partly to be invented, which concerns cities and which, except for the rationalism and modernism periods, characterized urban reorganization and reactivation of urban assets.

This approach has allowed the city of Amsterdam to develop *Amsterdam Circular Strategy 2020–2025* (City of Amsterdam, 2022), a strategy aimed to significantly reduce the use of new raw materials, thus contributing to a sustainable city. In the coming years, the city will map out various material flows, from entry to processing, to preserve valuable raw materials. The aim is to halve the use of new raw materials by 2030 and to achieve a fully circular city by 2050. The strategy aims to work in three values chains:

1. **Food and Organic Waste Streams**. Reduce food waste, not only in families, but also in hotels and restaurants; increase the consumption of regionally produced and plant-based food; properly process the waste stream and preserve the valuable, increasingly scarce fertilisers.
2. **Consumer Goods**. This value chain mainly concerns electronics, textiles, and furniture, goods that can often be repaired by professionals, and contribute to the depletion of rare raw materials.
3. **Built Environment**. This value chain is aimed at the organization of public space and the built environment: reusing building materials, adaptive reuse and improvement of existing buildings, sustainable building materials in public spaces, development of a marketplace for reusing building materials, and climate-adaptive solutions.

The approach proposed by the city of Amsterdam regarding the Built Environment, in perfect harmony with the *R-strategy*, unfortunately does not include the presence of mechanisms aimed at the co-creation of solutions for the reuse of unused or underused assets. Are envisage the development of a value chain only for the physical improvement and not a value chain as regards the contents of the integrated regeneration: adaptive reuse (functions), governance of the regeneration (involvement of different urban actors), and measurement of the expected impacts (social, cultural, economic, etc.).

2 The Co-city Mechanism: A City Capable of Addressing Reuse of Urban Commons

The urban model proposed in the Amsterdam strategy makes it possible to redevelop the "containers", buildings, and open spaces of the city thanks to a circular model, but we must also be able to reactivate a sense of community around these objects that otherwise they would remain outside the collective memory. It is therefore important to reactivate underused places within the urban fabric through collaborative processes because very often different values overlap in these "sleeping giants", such as:

- they might constitute an identity for one district or city sector, for example, former factories, empty military barracks, former industrial buildings, etc.
- they might be characterised by a high value of architecture or historical meaning.
- they might be of some interest to local communities to enhance the quality of life and of cultural services in the neighbourhood.

To be able to handle this complexity of "urban objects" and overlapping of values we must develop new management of these assets through bottom-up/collaborative approaches (co-creation actions) that can help local actors and urban authorities enhance their built heritage, strengthen local identity, and contribute to the preservation and/or reactivation of these assets. Despite their relevance, these kinds of practices, supported by different types of activities such as co-creation, participatory activities, temporary use within the assets, etc., are still experimental, and there is no systematic application because even if local administrations recognise the added value of such collaborative processes, promoting such practices is not easy for many reasons, mainly:

- Low project management of co-creation activities (insufficient human resources and adequate skills).
- Unclear local urban regulation (legal framework) to promote the cooperative management of the assets.
- Complexity of the urban asset's properties framework (patchy ownership).
- Difficult harmonization of local interests about the reuse of the built/cultural heritage.
- Lack of financial resources to promote "experimentation" of reuse (temporary) toward the new layout (functions and shape) of the assets.

In addition to these specific problems, related to the collaborative processes, challenges during the implementation of the regeneration projects might be related also to:

- A public procurement model not able to involve bodies from civil society (groups of active citizens, associations, etc.) that in some cases have introduced the idea of the re-use of one specific urban asset.
- Inconsistency and/or overlap of several regulations applying to the same place (building or open space) as Cultural Heritage protection, building requirements

as well as, environmental requirements that causing uncertainty in the regeneration process.

- Overlap of agreements and permits. In many states rehabilitation of ancient buildings must pass also through specific permits from the National Cultural Ministries. Managing such requests is for professionals and it is not easy, as a result, the timing of rehabilitation projects can be extended and being not feasible.

2.1 Keys Aspects for the Re-use: Dealing with 3 Typical Challenges

Starting from the reference framework described above the challenges that urban authorities must overcome to direct the city toward a co-city model, a city able to re-use their urban commons, and strengthen the concept of the Circular City, are essentially three:

1. Establish urban regulation encouraging the re-use of urban commons involving local stakeholders.
2. Include temporary uses of the buildings in urban planning tools.
3. Strengthen the competencies of local stakeholders, public servants, and technicians to co-design common rehabilitation and re-use strategies.

Encourage the re-use. To promote the re-use of buildings and spaces through collaborative management, it is key ensuring clear and transparent process. Therefore, it is important for the public authorities to: (i) build community awareness around potential opportunities; (ii) have in place the appropriate legal framework (urban regulation) to permit the involvement of local stakeholders (i.e., private sectors, civil society, research centre, university, etc.). The presence of a local urban regulation supporting the re-use of public assets (building and/or open spaces) is strictly interlinked to the assessment of their status. Urban authorities should keep track of dismissed/underused assets (and their state of play) so to be able to promote their rehabilitation processes answering to the following questions: (i) which functions and services are in needs in that specific area? (ii) who is to be involved and should participate in the various phases? (iii) how to promote the submission of ideas and participation and to deal with the results? (iv) which are the administrative acts (i.e., agreement, pact, contract, etc.) to be drafted among the urban authority and the local actors involved?

Temporary use triggers the re-use. The temporary use represents a means to explore the potential of places and it can become a task to aggregate the energy of local community around a project. To make the most of this potential, the re-use project must find the best mutual adaptation between use, users, and spaces. This brings to the core aspect for a successful collaborative project: the necessity to include temporary uses in the urban planning rules. Local planning legislation should allow for new temporarily functions and activities (even if not originally envisaged by the land-use rules).

New competencies to lead the re-use of urban commons. The physical redevelopment of urban dismissed, or underused, buildings or spaces through collaborative management processes forces public authorities to work by a result-oriented approach rather than by competences: functions and activities are defined through interactions and are not easily predetermined.

In summary, the co-creation process has at least two phases: "the external one" to involve and manage exchanges with external stakeholders (possibly from different sectors and of a significant large number); "the internal one" for the urban authority to be capable to support the ideas born in the previous phase. The first step implies abilities and competences of mediation and animation which are new for the public sector, while the second phase implies the capacity of working in a horizontal manner (disregarding the single sectoral competences) and therefore new internal procedures.

To be able to respond to this challenge in some European urban authorities, most Italian, a new role for the civil servant has been introduced: the so called "neighbourhood architect" or "community mediator". This is a role with new competences and functions, specifically designed to gear complex processes of regeneration and reuse of urban commons:

- **Enabler of local networks**: the hinge between local practitioners and the city's administration. It is a pivotal role for (i) analysing the practices on the ground to detect the relevant representatives and actors to involve; (ii) translating the wishes and needs of the parties involved; (iii) becoming the reference point for the interactive processes and its mediation.
- **Enabler of social innovation**: a key mediator able to organise and chair meetings and seminars (physical or digital) with interactive mechanism designed vis-à-vis the actors involved so to promote dialogue and facilitate common solutions/ agreements. It is a strategic role to manage co-design activities not only to gather local needs, ideas, and proposals, but also to identify possible common solutions and translate them into operational projects.
- **Enabler of technical aspects**: a key translator able to dialogue with the technicians, the private sectors (including also banking or public donors) and the local actors to transmit/translate ideas into projects facilitating the proposals respecting the format, the bureaucratic procedures, eligibility and timeframe and all other possible limits and further requirements.

2.2 *Key Activities for the Reactivation and Adaptive Re-use*

Starting from several best practices, developed in different European cities and aimed to facilitate the adaptive re-use of dismissed or underused buildings/spaces through collaborative management processes, we are able to propose a "toolbox" (Lanzoni, 2021), or a sequence of actions, that could be considered as the most relevant steps for the re-adaptation and enhancement of urban commons. We consider that mapping, a fundamental action to identify the urban assets, must be carried out by a local urban authority. The following phases, concerning the planning of reuse, can be carried

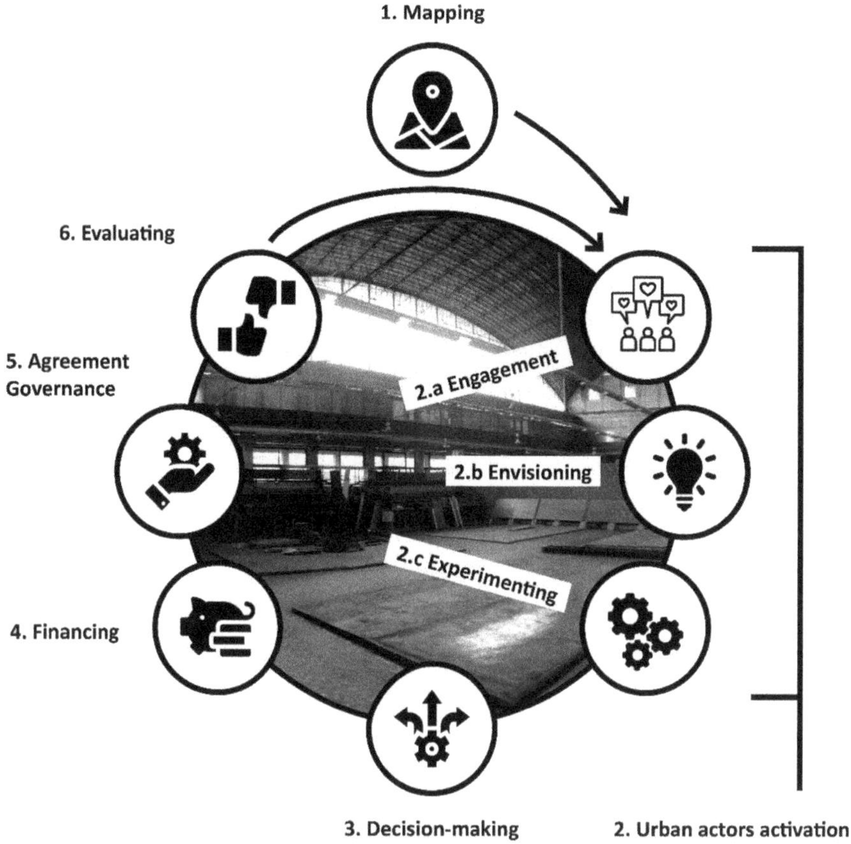

Fig. 1 The circular planning proposed for the management of the reactivation of urban commons (*Source* Chart taken by the authors)

out by the urban authority itself or by other urban actors, then continue over time to improve the management of the regenerated urban commons (Fig. 1).

1. **Mapping**. This is an important activity devoted to the identification of dismissed, underused, downgraded urban areas, sites, or buildings and to their analysis. These assets are those that worthwhile to be rehabilitated through collaborative and collective processes as they are of some value for the local identity. Thanks to the collective engagement and management, projects enhance social and cultural perspectives and services.[1] Mapping is drafted through innovative actions such as the community collaboration to highlight dismissed but relevant assets, to point

[1] A good practice for this topic is Longford NUA project, a community mapping project aimed to improve citizen engagement in the operation and service delivery of the Local Authority (https://www.longfordcoco.ie/longfordcoco/services/regeneration/longford-nua/).

out important practices carried out by civil society (groups of citizens, associations, etc.) using interdisciplinary approaches based on the civic collaboration (i.e., sociological, economic, cultural, environmental) and data management via geographic interfaces (Web-GIS).

2. **Urban actors' activation**. This paradigm includes different aspects, all integrated with each other, aimed at activating citizens with respect to the regeneration of spaces or buildings; co-designing reuse solutions; putting into practice small projects to test large redevelopments.

   2.a **Stakeholder's engagement**. The activation of the territory toward the challenge of the urban common's re-use, promoting matchmaking activities to change an inert situation. Raising awareness on the opportunities of underused spaces, on the need for a shared governance vision, supporting a co-design process, drafting an engagement action plan for the re-use are all possible activities that innovative Municipalities need to promote and enhance.[2]

   2.b **Envisioning and aligning the city challenges**. The drafting of scenarios, visions, common objectives, and all other possible inspiring tools are part of the engagement paradigm. Devoted to the development of a collaborative co-design activity, envisioning a common result is a mix of bottom-up and top-down activities that should be part (or are to be coherent) with the city's strategic planning systems.[3]

   2.c **Experimenting**. Testing proposals for the re-use of buildings and places identified through a collaborative process with temporary use can be a relevant action (implying a small investment) ensuring that the final project will be adequate, strategic, and sustainable. To better understand the essence of these points, compare the experience of the ZZZ Agency—ZwischenZeitZentrale[4] in Bremen (Germany).

3. **Decision-making**. The result of the engagement processes is to be coherent with the city's strategic planning systems. The decision-making step is addressed to improve the alternatives identified (and possibly tested) through vertical and horizontal integrated activities: vertical is about the coherence among the various

[2] A good practice for this topic is the *Murcia ADN urbano* project aimed at involving the citizens in the reactivation of three districts of the city (https://adnmurcia.es).

[3] A good practice for this topic is the experience led within the *Grands Voisins*, a transitional urban planning program, useful for identifying the principles for redesigning the area of a former hospital in Paris (https://lesgrandsvoisins.org).

[4] The ZZZ is a temporary use agency that works as an intermediate organisation between users, owners and the urban authority, and supports and initiates temporary use projects. The agency is supported and jointly financed by four departments of the city of Bremen (economy, construction, finances and culture). The ZZZ is usually involved in the development of concepts and strategies for deprived neighborhoods. ZZZ is considered as a way of promoting initiatives in the field of culture and cultural and creative industries, also in the field of social and environmental projects. Furthermore, temporary uses have been incorporated into the daily practice of the administration (construction law, contract drafting, guidelines for marketing, etc.) https://www.zzz-bremen.de/ueber-uns/.

project/strategic levels (regional, metropolitan, local, etc.) to frame the re-use of urban commons in a large strategic vision; horizontal is about the creation of coherence among the various sectors/departments within one urban authority in order to be able to fully trigger the potentiality of the projects.[5]

4. **Financing**. Conceive the financing model of the intervention (initial investment, management, and maintenance, among others) considering the cost/benefit analysis, subsidy elements, and their justification. Check for different sources of finances: public, private, user fee, alternative sources such as crowdfunding, crowdsourcing, among others.[6]
5. **Governance—Agreement**. The importance of a clear and stable legal framework to delegate the management of an asset, to draft agreements/pacts/contracts for the use of the assets. The legal instruments and definitions of contract's details are key for establishing an appropriate collaborative management process. Direct assignment to local associations can be justified based on a clear positive social and economic impact.
6. **Evaluating**. Adaptive re-use of urban commons is the act of giving new use to obsolete, underused, or dismissed places (buildings or open spaces) through collaborative management with the civil society and the third sector. In the evaluation of these processes is key important to use the following indicators: Ex-ante evaluation. It is the evaluation of the whole proposal, from the socio-cultural, and economic benefits; On-going evaluation (follow-up/output indicators) to monitor whether the benefits of the re-use are generating the expected outputs envisaged at the beginning; Ex-post evaluation (indicators of results) to check the impacts produced by the reuse project through quantitative and qualitative indicators of results.[7]

3 Co-city: Implications of This Approach

As previously highlighted, the development of strategies for the reactivation and reuse of urban commons (places, buildings, open spaces), with the strong involvement of local actors, is an innovative and constantly changing practice. There is good knowledge regarding participatory techniques, addressed to different groups of urban actors, but we cannot say that there is a consolidated working methodology for the reactivation of urban assets. In this moment of experimentation, as regards the

[5] A good practice for this topic is the experience led by the urban authority of Cesena (Italy) about the reuse of public urban assets, *Avanti, c'è spazio* project (https://www.comune.cesena.fc.it/avanticespazio).

[6] A good practice for this topic is the experience led by the urban authority of Melgaço (Portugal), Pop-up Shop project about the reactivation of the vacant shops within the city centre (https://vimeo.com/513344638).

[7] A good practice for this topic is the experience led by the urban authority of Chemnitz (Germany) within the ALT/BAU project aimed to the reuse of vacant residential buildings in city centre (https://www.alt-bau.eu/en).

regeneration of the urban assets (urban commons), and consequently also of urban sectors of the cities, we think that for the urban authorities it is useful to consolidate their knowledge in the following dimensions: a formal local legislative framework to promote the reuse; a continuous mapping of the territory to identify assets and practices; the ability to create harmony between the various existing strategic and planning tools; new professionals, capable of mediating the needs of local communities, their ideas, and the needs of local urban authorities.

Formal legislative framework providing clear, stable, transparent process. Legislative framework fostering and facilitating such transformative processes through collaborative management steps. It is highly recommended to draft normative references that favour the practice of reuse of the urban commons as a normal strategy for the urban planning management of the city. This means a legislative framework recognizing practices of temporary re-use and delegation of services to the civil society or the third sectors as a normal tool for the public urban authority to use and a clear and transparent procedures to involve the local stakeholders in the participative process. Reuse practices, supported by a clear legislative framework are successful, as in the case of the *Bologna Regulation on public collaboration for urban commons* (de Nictolis & Iaione, 2021).

Mapping of tangible assets and intangible practices and Atlas of practices. The identification of dismissed or underused areas (buildings and open spaces) is a fundamental step to better understand all possible details of the target places (i.e., location, state of play, interests, etc.). The mapping can be achieved through several different techniques: direct participative involvement, digital interactive tools, data for managing geographic information, etc. The mapping step can also be used as a starting point to engage the local stakeholders in actions of reuse. It is important to discover the abandoned/dismissed places of some value for the local communities, but also to create "a living atlas" of social, economic, cultural practices on-going in the city. Practices of local stakeholders (i.e., associations, NGOs, social profit and non-profit sectors, citizen, etc.) that manage spaces and services, giving values to places. The added value of having a global picture of the active positive practices for the public authority is to smooth the initial step of the stakeholder engagement, among other.

Creating coherence between the functions for the re-use proposed by local communities and the urban planning tools. The rehabilitation and the management of urban assets should be in line with services needed by the local communities and conceived in line with them: planning tools (Urban Agenda, urban Master Plan, Urban Strategy, Zoning, etc.) should reflect the local (often changing) needs while respecting the global challenges of ecological and digital transformation.

Developing involvement models (participation and co-design) and new professions: Agent of Local Development, Architect of the Neighbourhood, Community Mediator. Identifying the needs of the communities that live around the urban commons to be reused and evaluate solutions/projects with technicians of the urban authority for the implementation of the solution identified. In some cases, participation and co-planning activities have been positively guided by new professional profiles[8] (civil servant or external professionals representing the municipality) working as a hinge between the neighbourhood and the Municipality and among all the active bodies of the practice. These professionals (or other enablers) are needed to manage the entire co-design process, from the collection of needs to the definition of projects, guaranteeing a constant relationship among those involved. Their operational functions are, among others: detect/create the local network and keep it alive, collect ideas and mediate among different proposals, facilitate a common vision, develop the projects around the collected needs and common proposals.

References

City of Amsterdam. (2022). *Policy: Circular economy*. Retrieved from Amsterdam Circular Strategy 2020–2025: https://www.amsterdam.nl/en/policy/sustainability/circular-economy/

de Nictolis, E., & Iaione, C. (2021, June 12). The city as a commons reloaded: From the urban commons to co-cities empirical evidence on the Bologna Regulation. *The Cambridge Handbook of Commons Research Innovation, 24*. https://doi.org/10.2139/ssrn.3865774

Ellen MacArthur Foundation. (2015). *Growth within: A Circular Economy vision for a competitive Europe*.

Ellen MacArthur Foundation. (2017). *Cities in the Circular Economy—An initial exploration*.

European Commission—COM (2015) 614/2. (2015). *Closing the loop EN—An EU action plan for the Circular Economy*. Commission to the European Parliament, the Council, the European Economic and Social Committee and the Committee of the Regions.

Foster, S., & Iaione, C. (2016). The city as a commons. *Yale Law & Policy Review, 34*(2), 281–349. http://www.jstor.org/stable/43920369

Lanzoni, L. (2021). *Toolkit for the re-use of dismissed or abandoned open spaces or buildings for socio-cultural purposes through active participation. Urban Agenda for the EU*. Ecorys. https://futurium.ec.europa.eu/en/urban-agenda/culturecultural-heritage/library/action-4-toolkit-re-use-dismissed-or-abandoned-open-spaces-or-buildings

Marin, J., & De Meulder, B. (2018). Interpreting circularity. Circular city representations concealing transition drivers. *Sustainability, 10*(5). https://doi.org/10.3390/su10051310

Potting, J., Hekkert, M., & Worrell, E. (2017). *Circular Economy: Measuring innovation in the product chain*. PBL Netherlands Assessment Agency.

Prendeville, S., Cherim, E., & Bocken, N. (2018). Circular cities: Mapping six cities in transition. *Environmental Innovation and Societal Transition, 26*, 171–194. https://doi.org/10.1016/j.eist.2017.03.002

[8] To train the internal staff of the administrations in the management of complex processes of urban regeneration, there is a European Erasmus + project which is working on the development of a curriculum, training course for "new professionals" of urban regeneration called FUTURE – From Urban To hUman Regeneration (2020-1-IT02-K203-079250). The project wants to create a training model, that will support students and young professionals to acquire a set of competencies making them capable of implementing urban regeneration processes by reducing distances between stakeholders and creating opportunities for local sustainable development, co-design and enhancement of the identity of a territory (https://futureregeneration.eu/).

Chapter 5
Community Participation of China's Urban Regeneration: A Case Study of Sanxue Historical District in Xi'an

Jiahui Ji and Tim Heath

Abstract In the past two decades, community participation has been increasingly playing a critical role in urban regeneration. It refers to as an important part for the design and implementation of the urban regeneration process. Due to the persistent top-down tendency, particularly in developing countries, it is critical to expand the channel for community participation in urban regeneration practises. For achieving an inclusive target and enhancing positive interactions among different stakeholders, in recent years, partly as a response to the criticism of high-centralized regeneration projects, the community are increasingly empowered in the process of regeneration to bridge other stakeholders including government, private sector, and professionals in China. Through an appropriate participatory method, people could have an opportunity to act in the regeneration process, and cooperate with other stakeholders to complete a sound, long-term, and local-based solutions to regeneration (Li et al. in Cities (Elsevier Ltd.) 102:102743, 2020). However, although community participation approach is launched, people's demands have failed to be satisfied due to the continuing marginalizing the community. In order to dissect the reasons, by taking Sanxue historical district in Xi'an as case study, this paper aims to elaborate a new community participation model–collaborative workshop; analyses how collaborate workshop operates and why it finally results to the ineffective of participation. Grounded in a qualitative research methodology, this paper is built upon data collected through archival research, interviews with local community as well as other stakeholders, and on-site observation.

J. Ji (✉) · T. Heath
Department of Architecture and Built Environment, University of Nottingham, Nottingham NG7 2RD, UK
e-mail: Jihui.Ji@nottingham.ac.uk

E. Mangi et al. (eds.), *Innovative Public Participation Practices for Sustainable Urban Regeneration*, Urban Sustainability, https://doi.org/10.1007/978-981-99-9595-0_5

1 Introduction

In China, as major cities have faced some issues such as congestion and environmental degradation related to rapid urban sprawl, urban regeneration has become a necessity to enhance the urban functions and structures. Especially historical district, ever since a lack of concern for cultural heritage resulted from an overemphasize of economic growth, has shown the accelerated decay, which imminently required regeneration to achieve high-quality development. Nowadays, the connotation of historical district regeneration has broadened to enhance the physical environment, economy, and social dimensions comprehensively (Yang et al., 2022). The emphasize of community participation is also increasingly highlighted to break the previous single-centred cooperation only among the government, experts, and private sectors, which is regarded not only as a necessary practice but also as a key to success. However, although community participation has been considered significant, proper and effective participation process has yet to be established and enhanced, and community participation at best is 'going through the ritualized actions' and not endowed with decision-making power (Ericson, 2006; Yung & Chan, 2011).

Xi'an is an ancient city with rich cultural heritages. As the central city in western part of China, its regional influence and economic radiance has been increasing. The rapid economic development of Xi'an has also brought pressure and high demands on the regeneration of historic districts in Xi'an. In the early 1990s, while the real estate industry boomed, the historical districts were destroyed in a destructive manner, which brought considerable economic benefits in the short term and caused the irreversible damage to cultural legacies at the same time. In the mid-1990s, Xi'an was guided by the mainstream regeneration idea of 'restoring the olds as the olds (*Xiu Jiu Ru Jiu*)', and some regeneration projects were carried out in a conservative manner: the original traditional houses, streets, and alleys should be integrally protected (Wang et al., 2019). However, after many years of development, new problems have emerged due to the unreconciling conflicts between observation and changing contemporary needs. These problems should make us rethink about historical district regeneration, as community participation is regarded as an effective way so as to make urban spaces more 'people-oriented' (Arciniegas & Janssen, 2012).

By taking Sanxue historical district as the case study, this paper explores the community participation through collaborative workshop in the process of historical district regeneration in Chinese context. Specifically, divided into several section, this paper begins by understanding the definition of community participation in historical district in China and highlighting the importance of community participation in historical district; then, it introduces the Sanxue historical district and collaborative workshop; thirdly, the current challenges of spontaneous regeneration in terms of living space, public space and commercial space are presented. Lastly, the issues and reasons are deeply elaborated. This paper attempts to answer the following questions: what are the characteristics of the collaborative workshop in historical district regeneration in Chinese context; What are the issues during the community participation?

And why is that? As community participation increasingly plays an important role in historical district regeneration, the corresponding policies and actions should be adjusted in accordance with particular institutional background, which is necessary for China and other developing countries.

Case study was chosen as an appropriate approach. Single case study of Sanxue historical district cannot reflect all the issues regarding community participation of urban regeneration in China, but can depict details of collaborative workshop, and analyse the encountered issues and difficulties in community participation. Findings are based on the combination of secondary and primary sources. Secondary sources used to inform this paper are on diverse media sources: internet, journal papers, newspapers, and government policy documents. Furthermore, the major issues, challenges, and reasons of ineffectiveness in community participation are presented and identified, which were verified by 7 semi-structure interviews.

2 Community Participation of Historical District Regeneration in China Context

World bank defines public participation as "a mechanism by which people can exert influence over the entire process, including policy formulation, design alternatives, management, and monitoring of development interventions in the communities" (The World Bank, 1992, p. 2). Since it originated from the British planning system in 1947, public participation is also seen as an indispensable component in the field of planning (Taylor, 1998). In order to produce a scientific and reasonable urban planning and decision-making, a joint system among all stakeholders, including the local government, private sectors, and community is established, with the empowerment allocated to the public to foster their expression of thoughts (Li et al., 2020).

As for urban regeneration in historical district, community participation can be utilised for alleviating the conflicts of the conservation and regeneration in which community, government, and experts' usually hold different views (Ericson, 2006; Yang et al., 2022). In the words of Washington Charter of 1987: "the participation and cooperation of the citizens are crucial for the success of the conservation programme and should be promoted" (ICOMOS, 1987). In cultural heritage conservation and regeneration, several trends have evolved and made community participation an important element of the decision-making process: First, the heritage value has appeared a "people-oriented" shift rather than a mere "culture-oriented"; second, grassroots support has replaced top-down support (Yung & Chan, 2011). It is argued that heritage regeneration by relying on public perceptions rather than elites' judgement can make a real difference (Su & Wall, 2014; Tweed & Sutherland, 2007).

In west countries, community participation has a long history. China, on the other hand, differs from the western countries in causes of institutional structures, so it is necessary to realize how the background in China look like. The community development in China has been characterized by top-down leadership (Zhai & Ng, 2013;

Zhang & Fang, 2004). In order to promote administrative efficiency, the government usually takes more responsibilities to influence community development through policies, guidance, and other measures (Li, 2006; Li et al., 2019). In other words, the government always dominants almost everything for a long time. Thus when the developing countries is delved into in greater depth, it is important to put more attention to comprehend the adaptation and specialization in community participation against "institutional background". That is to say: community participation method applied in China under the circumstances of high-centralized authority may lead to the results that prioritize the government's desires and official agenda.

3 Study Area and Overview of the Stages of Development

Sanxue historical district is one of Xi'an's conservation areas, bordered by the ancient walls. Surrounded by several historic temples and traditional houses registered for protection by the Xi'an, Sanxue covers an area of approximately 12.2 hm^2 (Zhang et al., 2014). The stages of development of Sanxue historical district can be divided as followed (Figs. 1 and 2).

Tang Dynasty—Yuan Dynasty

Sanxue historical district belongs to Beilin block, and its origin can be traced back to the Tang Dynasty (Xi'an Beilin District Chorography Committee, 2008). Sanxue historical district is a part of imperial academy, the earliest central institution of higher learning in China. During Song Dynasty, many steles such as famous 'Shitai Xiaojing (石台孝经)' were moved to the Forest of Steles, which significantly laid the cultural foundation. In the second year of Chongning in the Northern Song Dynasty, the government school of Jingzhao and the Confucian temple were relocated and reconstructed, forming the basic pattern consisting of 'government school-Confucian temples—Forest of Steles' in Sanxue historical district (Ibid.). Jv Si (local educational institutions), Gong Yuan (the institution that holds the provincial examination in Shaanxi province) and the highest educational government schools successively appeared in the Sanxue historical district, thus establishing the status of cultural and educational centre in Xi'an (Ibid.). At that time, the functional attributes of Sanxue historical district became a more complex compound block.

Ming Dynasty—The Republic of China

In Ming Dynasty, Guanzhong Academy (Guan Zhong Shu Yuan) was built on the west side of Sanxue historical district, so a concentrated cultural and educational area began to form (Zhang et al., 2014). When the scale was increased in Qing Dynasty, its cultural radiation to the city strengthened, and at the same time attracted a large number of scholars engaged in calligraphy, epigraphy and history to gather around, forming a unique cultural tradition (Jiang & Su, 2019). However, at the beginning of twentieth century, the profound social changes violently impacted the traditional culture mainly centred on Confucianism (Ibid.). Especially Sanxue historical district,

Fig. 1 The location of Sanxue historical district (*Source* By the authors)

which directly served Confucianism and old-style education, bears the brunt. It was not until the Republic of China that Sanxue Street, as a cultural and educational centre, got more attention again (Xi'an Beilin District Chorography Committee, 2008). At this time, the culture attribute of Sanxue historical district recovered and rebecame a communication place for many literati.

After 2000—A Period of Spontaneous Construction by Residents

The development after 2005 marked an early stage of spontaneous regeneration from local people. The physical environment in Sanxue historical District has been in backward situation, especially in the background of rapid urbanization in Xi'an, which is in sharp contrast with the rapid development of the surrounding areas. In addition, the influx of floating population also adds more heavy burden to the existing old infrastructure. Faced with the dilapidated living conditions in Sanxue historical

Fig. 2 Main street of Sanxue historical district (*Source* By the authors)

district, through the investigation of residents' willingness to launch regeneration activities, the survey data shows that the residents were very eager to carry out urban renewal projects, hoping to improve their living environment and local public facilities (Chen, 2017). Therefore, local people' self-building behaviours are constantly appearing. By adding floors and additional components, expanding building areas, and other self-organizing construction methods, the available space is increased, and the building density and floor area ratio in the block are gradually increased. By comparing the data in Sanxue Street area in 2006 and 2016, after ten years of spontaneous regeneration from local people, the building density and plot ratio have obviously improved (Ibid.).

The Sanxue historical district in Xi'an is a significant cultural, social, and economic asset and one of the two finest preserved historic districts. However, a terrible living environment and a dwindling sense of community are generated for locals when rapid urbanisation and population expansion have resulted in uncontrolled construction, poor infrastructure, deteriorating building stock, and subpar community service facilities (Ibid.). Fortunately, a collaborative regeneration workshops involved in community participation has been proposed, which aims to regenerate the Sanxue historical district through local people's efforts.

4 Proposing Collaborative Regeneration Workshops for Community Participation and Its Results

Through the workshop, the community can communicate with other stakeholders including government, experts, and developers etc. In this way, all stakeholders can take co-actions to create feasible solutions to regenerate Sanxue historical district. Nevertheless, after the implementation, it turns out that the collaborative workshop cannot reach its full potential, and irregular spontaneous regeneration actions still exist.

4.1 Proposing Collaborative Regeneration Workshop

Although the top-down forces, such as government and developers, play a vital role in the urban regeneration and can greatly influence urban space, the residents' satisfaction cannot be neglected in the decision-making. In response to this, a shift to an integrated people-centred approach on historical district regeneration has come to the fore in China's current urban planning practices. This collaborative workshop consists of four stages: consultation and negotiation, collaborative planning and design, applying for construction approvement, and officially permitted construction. This workshop focuses on the inputs of people, and claims that it is a spontaneous regeneration process mainly dominated by the residents.

Stage One: Consultation and Negotiation

In order to address the problem of promoting public participation and social justice, a 'collaborative regeneration workshop' was set up with the support of the government, inviting professional planners, community organizations, and residents to come together. This platform thus creates a platform for dialogue between multiple stakeholders to reconcile conflicting interests. The initial stage is consultation and negotiation, which should be organised by community committee and aims to let the residents know the preliminary understanding of Sanxue historical district, including historical remains, population, industry etc. Based on the preliminary understanding, some space problems and the ways to protect and activate the historical and cultural resources through regeneration in Sanxue historical district are elaborated. The residents are encouraged to share their viewpoints of future development.

Stage Two: Co-planning and Design

The second stage of workshop is co-planning and design. The collaborative workshop not only mobilizes the residents to express their views, but also should consult with the government (Beilin district urban planning bureau and local street office), as well as design institute, such as Xi'an urban planning research institute, and other universities, assist in design and scheme, and finally deliver and publicize a working report and feedback of the residents. Some community activities can also be held regularly such as seminar, community salon in reception room or public square. Through these co-decision-making activities, the attitudes of residents regarding community regeneration projects might progressively change, moving from passive participation to more active in communicating with planners and government. In the workshop, all stakeholders concerned can come to an agreement on how the regeneration should be carried out, so the related regeneration scheme can be designed smoothly.

Stage Three: Applying for Construction Approvement

This stage is an important part of workshop. After reaching consensus with other stakeholders, obtaining the construction planning permit is the legal pre-requisite for carrying out construction. The owner of the residential property is required to prepare the relevant documents, such as formal application, for the preliminary examination by the street office and planning department. Once all the required documents are examined, the owner will then submit all the construction documents (architectural design drawings with official seal should be provided) to the Beilin branch of the Xi'an municipal planning bureau for the planning permission procedure. Finally the planning branch grants the 'construction project planning permit' within the prescribed time to owners. During the period of obtaining construction project planning permit, the collaborative workshop also offers technical assistance, including expert design, for the general public. As a result, it should be mentioned that the workshop performs an organising and coordinating role. First, residents may benefit from the collaborative workshop with research groups in preparing architectural design and construction drawings. On the other hand, the government may receive the condensed information which may occasionally act as a catalyst for encouraging policy support.

Stage Four: Officially Permitted Construction

After the construction permit is issued by the planning bureau, the owner of the residential property can apply for the construction permit at the construction bureau, and then the construction can be carried out according to the declared construction project and relevant requirements, and the construction has to be started within 6 months from the date of obtaining the construction permit. If, during the construction process, there is a need to make changes to the architectural design scheme (including partial adjustment of building functions), an application for change or reapproval must be submitted to the Xi'an municipal planning bureau. The workshop is also responsible for the daily supervision and management of the approved residential construction projects. If the illegal construction actions occur during the construction of the residential building, including the changes of external dimensions, height, shape, and other construction acts affecting urban planning, or failure to implement promised supporting facilities in accordance with the requirements of the permit, the workshop is responsible to call halt to and order to correct the behaviours.

4.2 Current Challenge: Constant Actions of Spontaneous Space Regeneration Centred on Community Residents' Needs

After teasing out the process of collaborative regeneration workshop for community participation, By providing residents a variety of means of bargaining and negotiation, compromise, and of course, participating in the process, Sanxue historical district appears to attempt to embrace an inclusive participation strategy. However, the results of field work have shown that the residents generally disagree with the mode of collaborative workshop, so the spontaneous space regeneration mainly in terms of residence, commercial and public space are continuing.

Residence space: Constantly changing lifestyle including the need for economic growth and quality of life have led to residence space changes. It can be seen from the investigation that small-scale additional components are the most common form of self-organised residence space regeneration in Sanxue historical district. This type of regeneration is universal, efficient and simple to implement, and is accepted and adopted by most residents. According to research statistics, there are 79 residential properties in the Sanxue historical district that have experienced small-scale additional components construction (Chen, 2017). During the investigation, many residents have added one more storey used as storage rooms or rented as guest rooms to the original property (Fig. 3).

Public space: The spontaneous public space regeneration in Sanxue historical district is mainly through the informal and gradual occupation, which has given rise to a variety of non-standardised design and mainly embodies in extending private space. Firstly, the main operation is to enclose the edge space by building canopies to form a semi-public space, thereby privatising the public space and expanding the business area. This grey space formed by the additional canopies is both a public place and a private place for the businesses, presenting the characteristics of half public and half private, half open and half closed. Secondly, some shops often move

Fig. 3 The added components of residence space (*Source* By the authors)

Fig. 4 The occupied public space (*Source* By the authors)

their items to pavement during business hours, reducing the distance to pedestrians and attracting potential customers (Fig. 4).

Commercial space: The commercial spaces that belong to spontaneous regeneration mainly refer to the direct conversion of other functional space into commercial space within Sanxue historical district. This type of conversion is usually found on the ground floor, as it is more convenient for many landlords to replace original spaces on the street side with commercial space to increase the rental fees. Generally, this conversion is illegal in Sanxue historical district because of the change of attribute, however, driven by economic benefits, residents still are wild about conversing the space function. For instance, in the 106 Dongmutou courtyard, the owner has turned the residence space of the two rooms along street into commercial space in order to increase rental income, and the public traffic space adjacent to the courtyard has also been used as commercial space for a certain period of time. This spontaneous commercial space regeneration lacks formal planning, but is a way for the owner to increase the rent of the house significantly with minimal construction costs (Fig. 5).

5 The Summary of Issues of Community Participation in the Sanxue Historical District

The following difficulties are revealed via in-depth case study analysis and extensive sources. It was found that these raised issues are commonly agreed by stakeholders after additional discussion with the various stakeholders in a series of interviews.

Fig. 5 Conversion into commercial space (*Source* By the authors)

5.1 Conflicts of Interests

In broad term, as the government and residents are making up two major groups of stakeholders in Sanxue historical district, the interests related to these two stakeholders can directly affect the effectiveness of public participation activities (Yung & Chan, 2011). The government put more focus on the constructing and enforcing the cultural characteristics that can be regarded as sources and exploited to promote local economy, while the residents are concerned with improving quality of the living environment and income, and they also increasingly disagree the approach that favours the cultural heritage conservation in Sanxue historical district and also see heritage conservation as a barrier to hinder city development, so the cultural heritage should give way to development and cater to the contemporary needs. It can be clearly seen that the proposed vision of Sanxue historical district regeneration imposed by the government is centred around maintaining authentic architecture style and environment to promote traditional Chinese culture in order to enhance the uniqueness to generate local economy growth. Nevertheless, these objectives rarely take actual contemporary needs of the residents into account. Although locals may express their desires and concerns in workshops, the local government's preference for cultural and economic benefits makes it unlikely that their opinions will be listened. Regarding the vision of Sanxue historical district regeneration, as the leader of street office shared the opinion:

> Sanxue historical district is one of the key conservation units, so we should utilize the uniqueness and regenerate it to attract more businesses to settle in and promote local economic development. We also did some space beautification works, but they are distributed near major cultural legacies. (government leader 1).

In addition, different views are also frequently seen even among the local people when response to the regeneration of Sanxue historical district. During the interview, some residents said that the historical relics were out of date and should give way to development. For instance, an occupier held the opinion that the current living quality of local people should be placed as priority. On the other hand, a local person said the Sanxue historical district should be preserved because it was filled with past memories. These different interests of stakeholders probably resulted from their dissimilar background and further form the challenge in resolving these conflicts. As different residents expressed:

> Our living quality is much lower that the surrounding, so I don't think the historical building deserved to be protected because they have conflicts to our daily needs. We need normal life, so we have started some constructions to satisfy our needs (resident 2).

When we were young, we often play in the lanes, around historical buildings. One of my old friends has lived in an old house for many years. These our special memories make me feel it is my place (residents 4).

5.2 *Unequal Power Position*

Although the collaborative workshop aims to stress the community participation during regeneration projects, the negotiations between the residents, who are a relatively less powerful stakeholder, and the government, which has dominating power, are not on the basis of equality, reinforcing what Arnstein emphasises about the need for power to influence the outcome of public participation (Arnstein, 1969). In fact, the government still holds the prevailing power in urban planning and conservation that is not allocated to residents, which is likely ingrained in enduring top-down narratives. In the Sanxue historical district case, the government has the power to make decisions on whether or not and how to implement the regeneration constructions and can further launch a series of consultations. Since the decision ultimately lies with the government, its original plans can be forged ahead according to government's will, despite the inadequate consultation and marginalized various voices from local people. Even though a sound procedure of community participation has been formed, the community participation in Sanxue historical district mainly turns out at the degrees of tokenism (Wei, 2022), and the degrees of community power is eventually at a low level. One tenant in Sanxue historical district stated that they knew that the public opinions cannot influence government deeply to make the decision on regeneration projects:

> I know it is just a formalism in the first place. It is true that our opinions cannot influence the final results. In general, civil servants and government-appointed professionals are always elected powerful stakeholders and determine what would happen next. Since the collaborative workshop was launched, it was poorly rated indeed. Although a much greater willingness of community participation has shown, we still have a long way to go (resident 1).

Actually we really want the residents to join in and decide what to do next. But the reality is there are too many requirements, we cannot satisfy everyone and most importantly, we need to keep the project going through as expected (government leader 2).

5.3 Non-transparent and Ambiguous System

In the case study, this collaborative workshop has been a non-transparent and ambiguous system, which has confused the decision on what to do with the regeneration. Indeed, since local people's plans are different from each other, all the wishes of the community cannot be included into final blueprint. Whenever a series of public forums and consultations, held by workshop, aiming at gathering views from the community are launched, conflicts still exist because regeneration planning is a very complicated process involved a wide range of people with different appeals. Hence, it is a difficult task to generate a plan, which everybody will agree on. Ultimately, the collaborative workshops including consultation, negotiation, and plan inspections are not a platform to realize democracy, but rather as a way to smoothening and simplifying the participation process through superficial community participation, with the purpose to limiting the empowerment on local people. Some residents also complained:

> Most residents living here still don't know clearly about the procedure. When we asked the office leaders, the answers we got were hard to understand, and not to mention how to carry on (Resident 3).
>
> There are so many complicated details about the regeneration projects. Actually, I couldn't fully understand every detail. I think the government can guide and support us to complete it, rather than just give us some misunderstanding information. (Resident 5).

6 Discussion and Conclusion

In the case study of Sanxue historical district, the government initially launch the regeneration project through workshop for attempting to resonate spontaneous construction and further improve living environment. Undoubtedly, with the organization from the government, China's community regeneration projects have been shown a united and complete procedure, but there are still a series of difficulties encountered during the Sanxue historical district regeneration project. It turns out that unreasonable spontaneous regeneration constructions still happen and residents in the projects had a low-level of satisfaction and lack a sense of belonging and cohesion, corresponding to some research findings that residences often hold negative perceptions of community participation.

According to the analysis results, some reflections on community participation are elaborated as followed. Firstly, other than identifying the divergent requirements and perceptions of different stakeholders, balancing and resolving the opposing interests of stakeholders through an appropriate analysis is of great significance for community participation (Yung & Chan, 2011). Sanxue historical district failed to have positive consequences partly due to the putting more focus on political responsibility through authority's motivation, while the current communities' interests are neglected. This type of community participation is to some certain a 'hegemonic' device, aiming to maintain and secure the existing power structures. In other words, taking the form of communicating with and seeking commitment of residents and then allowing 'responsible autonomy' within limits is superficial democracy and a more gentle and subtle tyranny, because the essence of collaborative workshop in Sanxue historical district is implementing the projects more efficiently according to government's will rather than seeking 'real' participation, which ends up reinforcing the benefits and power of governments. Secondly, even though community involvement in planning is increasingly acknowledged as a crucial component of planning, this recognition has largely been verbal and has not been converted into action, particularly in developing nations where citizens have not shared equal authority. The local government continues to overcarry the burden of organising, making decisions, and providing guidance, despite the fact that some efficient communication channels between the government and residents have been suggested (Li et al., 2020). As a result, discussions and negotiations on the interests of community residents are merely "somewhat going through the motions". Due to the disadvantage that community engagement is limited to a planned official process to serve a predetermined conclusion, all pre-planned by the authorities, the community participation typically ends with the superficial appearance and is an admittance to having become ritualised.

The community participation in a form of collaborative workshop in Sanxue historical district is not individual case in that regard, which would be clearly elaborated if related to China's long-term top-down leadership. Some scholars use 'hegemony' to describe and criticize this phenomenon that Chinese government undertakes a more specific and comprehensive responsibility (Wei, 2022). The historical regeneration and empowerment of community participation are fundamentally conflictual because government-community relations may be better perceived as inevitable structural problems. This unequal and stubborn relations resulted from political background, in various ways, have the potential to politicise what ought to be a process of deliberate depoliticization (Hickey & Mohan, 2004; Wei, 2022). Based on this case study, it is insightful to understand and reflect upon what is effective participation, and how the unbalanced and reinforced power exclude marginalised groups based on the critique to this unequal participation.

Regarding the particular methods of the collaborative workshop, the Sanxue Historical District Case Study is not the only historical districts that adopt it in China. However, the defective participation system and the detrimental effects of community engagement can also be seen helpful experiences. To conclude, focusing on the pursuit of the authority rather than the interests of the community may lead to unresolved disputes and a strategy that is predetermined by the authority without community input. If the community could truly be empowered, it might be conceivable in practise to pursue a more favourable outcome, which calls for more thoughts and efforts spanning different disciplines in which scholars and practitioners can continue working.

References

Arciniegas, G., & Janssen, R. (2012). Spatial decision support for collaborative land use planning workshops. *Landscape and Urban Planning (Elsevier B.V.), 107*(3), 332–342. https://doi.org/10.1016/j.landurbplan.2012.06.004

Arnstein, S. R. (1969). A ladder of citizen participation. *Journal of the American Institute of Planners, 35*(4), 216–224.

Chen, Z. (2017). *Research on self-support requalification of residential living space in Xi'an Ming City Sanxue Street Area.* Xi'an University Architecture and Technology. Retrieved August 2022, from https://cdmd.cnki.com.cn/Article/CDMD-10703-1017737077.htm

Ericson, J. A. (2006). A participatory approach to conservation in the Calakmul Biosphere Reserve, Campeche, Mexico. *Landscape and Urban Planning (Elsevier B.V), 74*(3), 242–266. https://doi.org/10.1016/j.landurbplan.2004.09.006

Hickey, S., & Mohan, G. (2004). Towards participation as transformation: Critical themes and challenges. In S. Hickey & G. Mohan (Eds.), *Participation: From tyranny to transformation? Exploring new approaches to participation in development* (pp. 3–24). Zed Book.

ICOMOS (1987). *Charter for the Conservation of Historic Towns and Urban Areas* (Washington Charter 1987), Adopted by ICOMOS General Assembly in Washington, DC, October 1987. Available from: https://www.icomos.org/images/DOCUMENTS/Charters/towns_e.pdf

Jiang, M., & Su, W. (2019). Research on the planning strategy of Sanxue Street historic district in Xi'an from the perspective of interest balance. *Journal of Green Science and Technology, 12*, 259–269. https://doi.org/10.16663/j.cnki.lskj.2019.12.101

Li, W. (2006). Community decision-making participation in development. *Annals of Tourism Research (Elsevier Ltd.), 33*(1), 132–143. https://doi.org/10.1016/j.annals.2005.07.003

Li, X., Hui, E. C. M., Chen, T., Lang, W., & Guo, Y. (2019). From Habitat III to the new urbanization agenda in China: Seeing through the practices of the "three old renewals" in Guangzhou. *Land Use Policy (Elsevier Ltd.), 81*, 513–522. https://doi.org/10.1016/j.landusepol.2018.11.021

Li, X., Zhang, F., Hui, E. C., & Lang, W. (2020). Collaborative workshop and community participation: A new approach to urban regeneration in China. *Cities (Elsevier Ltd.), 102*, 102743. https://doi.org/10.1016/j.cities.2020.102743

Su, M. M., & Wall, G. (2014). Community participation in tourism at a world heritage site: Mutianyu Great Wall, Beijing, China. *The International Journal of Tourism Research (Blackwell Publishing Ltd.), 16*(2), 146–156. https://doi.org/10.1002/jtr.1909

Taylor, N. (1998). *Urban planning theory since 1945.* SAGE.

The World Bank. (1992). *Participatory development and the world bank: Potential directions for change.* Discussion paper number 183. The World Bank.

Tweed, C., & Sutherland, M. (2007). Built cultural heritage and sustainable urban development. *Landscape and Urban Planning (Elsevier), 83*(1), 62–69.

Wang, S., Yuan, J., Xu, Y., Zhang, L., Li, X., & Zhu, L. (2019). Sustainability of historical heritage: The conservation of the Xi'an City Wall. *Sustainability (MDPI AG), 11*(3), 740. https://doi.org/10.3390/su11030740

Wei, R. (2022). Tyrannical participation approaches in China's regeneration of urban heritage areas: A case study of Baitasi Historic District, Beijing. *International Journal of Heritage Studies (IJHS) (Routledge), 28*(3), 279–296. https://doi.org/10.1080/13527258.2021.1993310

Xi'an Beilin District Chorography Committee. (2008). *Xi'an Beilin District chorography.* Sanqin Press.

Yang, J., et al. (2022). Community participation strategy for sustainable urban regeneration in Xiamen, China. *Land (Basel), 11*(5), 600. https://doi.org/10.3390/land11050600

Yung, E. H. K., & Chan, E. H. W. (2011). Problem issues of public participation in built-heritage conservation: Two controversial cases in Hong Kong. *Habitat International (Elsevier Ltd.), 35*(3), 457–466. https://doi.org/10.1016/j.habitatint.2010.12.004

Zhai, B., & Ng, M. K. (2013). Urban regeneration and social capital in China: A case study of the Drum Tower Muslim District in Xi'an. *Cities (Elsevier Ltd.), 35,* 14–25. https://doi.org/10.1016/j.cities.2013.05.003

Zhang, Y., & Fang, K. (2004). Is history repeating itself? *Journal of Planning Education and Research (SAGE), 23*(3), 286–298. https://doi.org/10.1177/0739456X03261287

Zhang, R., Li, H. Y., Bi, L., & Cui, J. Q. (2014). Research on conservation and regeneration of historic district in the background of fast urbanization—Case study of Xi'an Sanxue Alley Historic District. *Applied Mechanics and Materials (Trans Tech Publications Ltd.),* 584–586, 407–414 (Architecture, Building Materials and Engineering Management IV). https://doi.org/10.4028/www.scientific.net/AMM.584-586.407

Part III
Participation in Informal Settlements Regeneration

Chapter 6
Improving Liveability Through Regenerating Informal Neighbourhoods: A Case of Old Dhaka, Bangladesh

Md. Nawrose Fatemi, Zareen Habiba Islam, and Tahmina Rahman

Abstract Dhaka, Bangladesh's capital, has been the world's most densely inhabited city since 2016, with a density of 30,093 people/km^2 in 2022, posing challenges in providing access to basic civic facilities to its residents. The planning history of Dhaka reveals that both planned and spontaneously developed neighbourhoods have evolved here throughout time. Besides these two distinct types of development, the historic older part of Dhaka with all its cultural and social heritage has expanded within its physical limit. Consequently, citizens' liveability conditions in Old Dhaka characterized by the low standard of urban services, narrow road networks, poor drainage systems and congested urban fabric, have steadily worsened over time. A new Detailed Area Plan (DAP) has recently been restructured and published to develop Dhaka into a better liveable city, resolving its urban challenges, ensuring citizens' equity proportionate to urban amenities, upgrading urban services, and gradually improving the physical, social, economic and environmental condition of different areas. Engaging locals during the fieldwork, this study attempts to identify the existing problems, their experience living with these challenges and the aspirations of the locality regarding the new urban regeneration proposal of DAP. Several participatory approaches were exercised, including focus group discussions, key informant interviews, transect walks and community mapping of the existing conditions and two urban design workshops were conducted with 50 participants to demonstrate the urban regeneration proposals of DAP for two Old Dhaka neighbourhoods. This paper suggests solutions in terms of policy and design to provide a liveable environment for the locals in terms of physical infrastructure, amenities and facilities while conserving the socio-cultural fabric of the neighbourhoods. This study concludes that balancing top-down and bottom-up approaches in design solutions can ensure liveability not only in terms of physical outcomes but also regenerate the embedded socio-cultural aspects of these neighbourhoods.

Md. N. Fatemi (✉) · Z. H. Islam · T. Rahman
Department of Architecture, University of Asia Pacific, Dhaka, Bangladesh
e-mail: nawrose@uap-bd.edu

E. Mangi et al. (eds.), *Innovative Public Participation Practices for Sustainable Urban Regeneration*, Urban Sustainability, https://doi.org/10.1007/978-981-99-9595-0_6

1 Introduction

People attract to cities due to the pull of a better life supported by proper public services and amenities, a better economy offered by diverse jobs, businesses and similar income options and improved communication facilitated by roads, streets and other infrastructures (Chen & Rosenthal, 2008; Glaeser, 2012). Thus, rapid urbanization and unprecedented economic growth have been observed in several metropolitan cities, especially in recent decades (Lee, 2021). Scholars on urban sciences (Dave, 2010; Hee, 2015; McGee, 2010) have underlined that high density offers issues of congested neighbourhoods, poor quality of urban services, and increased disputes over access to public facilities, implying the end of city liveability. As the world's most densely inhabited city since 2016, Dhaka has been tackling similar challenges in providing access to basic public facilities to its residents. The population density of Dhaka reached 30,093 persons per square kilometre in 2022 (SRD, 2022). Rapid natural population increase, as well as significant levels of interior migration spurred by greater economic opportunities and rural poverty, have led to Dhaka's population expanding at a 4% yearly pace during the last three decades (Ahmed, 2021).

Within its planning framework, two spatial patterns i.e. planned and spontaneously developed neighbourhoods are dominant in Dhaka throughout time (Islam & Fatemi, 2015). Besides these two distinct types of development, the historic older part of Dhaka with its entire cultural and social heritage has expanded within its physical limit. Therefore, the Old Dhaka area suffers from low standards of urban services, narrow road networks, poor drainage systems and congested urban fabric (Tokey et al., 2020). Consequently, citizens' liveability conditions in this area have steadily worsened over time (Iffat & Fatemi, 2012; Nahrin, 2018). The recently restructured and published Detailed Area Plan (DAP) 2022–2035 has a conceptual foundation that includes a statement in favour of social inclusiveness and well-being (Ashraf, 2022). To develop Dhaka into a better liveable city, DAP 2022–2035 tried to resolve its urban challenges, ensure citizens' equity proportionate to urban amenities, upgrade urban services, and gradually improve physical, social, economic and environmental conditions of different areas. Engaging locals of Old Dhaka during the fieldwork, this study focuses to identify the existing problems of the area, their experience living with these challenges and the socio-spatial qualities of the locality regarding the new urban regeneration proposal of DAP 2022–2035. This chapter also demonstrates a few solutions in terms of policy and design to provide a liveable environment for locals in terms of physical infrastructure, amenities and facilities while conserving the socio-cultural fabric of the neighbourhoods.

2 Background Knowledge

2.1 Liveability and Regeneration of Urban Neighbourhoods

Liveability denotes an area's living circumstances and represents a citizen's perspective of whether or not the neighbourhood is suitable for living (Fatemi & Islam, 2014). Thus, liveability supports place-making as well as sustainable development with a focus on the neighbourhood and its surroundings (Leh et al., 2020). Rahman and Fatemi (2021) emphasized the need for diverse density for different locales with civic facilities, and public amenities, as well as appropriately built and effectively connected public open spaces to ensure liveability. Badland et al. (2014) identified liveability as an index of the subjective well-being and quality of life of the locals. Consequently, communities' experiences have been utilized by urban researchers to evaluate neighbourhood quality (Leby & Hashim, 2010; Salaripour et al., 2021), because the human-built attributes of neighbourhoods have a substantial effect on their psychological and social well-being. Correspondingly, urban regeneration has the potential to reshape neighbourhoods, improve their social image, and create dynamic and appealing places that stimulate long-term inbound investment (Stafford & Baldwin, 2018). Oh et al. (2013) stressed the importance of community-based indicators when looking at liveability in connection to neighbourhood revitalisation, since the revitalisation process promotes sustainability, quality of life, people-centred development, and participation. Therefore, more participation from local communities is anticipated to provide a contextual understanding of liveability and end-user perspectives on neighbourhood revitalisation (Hankins & Powers, 2009; Woolcock, 2009). Moreover, neighbourhoods continue to be effective social and geographic units of common identity that serve as the foundation for communal initiatives to tackle diverse urban challenges (Looker, 2012). Community participation has become a common tool for neighbourhood revitalisation in the U.S. (Rock, 2018), Japan (Okyere et al., 2019) and other countries (Clark & Wise, 2018). This idea frames the participatory approach to neighbourhood revitalization as a formalised collective effort to assess and improve liveability, an attribute that might vary from one context to the next (Choi & Seo, 2020; Vine, 2012).

2.2 Regeneration Schemes of Old Dhaka

Throsby (2005) explained that the liveability of an urban environment can be characterized not only by tangible attributes, e.g. public infrastructure availability, but also by intangible features, e.g. local identity, memory and image of the place and social networks. Correspondingly, Said et al. (2014) indicated social and cultural indicators, townscape improvement indicators, economic regeneration indicators, and image and identity indicators as key indicators for heritage-led regeneration procedures. If the city officials consider the regeneration of Old Dhaka, all of these four indicators are valid. However, it is also necessary to assure that this regeneration procedure respects the local community's hopes and expectations (Akter et al., 2017). Dhaka has developed from a tiny trade town to a megacity during its 400-year history (Iffat & Fatemi, 2012). The existing spontaneous spatial character of Old Dhaka has grown with generations representing the lifestyle and people of this area. This indigenous settlement is defined by intricate paths and lanes that run through the neighbourhoods, where shophouses are a prevalent architectural style (Islam, 2016). Residents of Old Dhaka have a long tradition of socializing with one another in public settings (Islam & Fatemi, 2015). Traditional hierarchical places i.e. Gali (lane), Morh (node), Mahalla (neighbourhood) and Chowk (Square) were formed as a result, which were all sequentially developed along the streets (Mowla, 2004). Thus, these streets have transformed into 'places' for a variety of rituals and festivals, while the shopfronts of the buildings have served as 'platforms' for social interaction. Additionally, houses with courtyards (Uthan) and small terraces on different levels are another common architectural style in Old Dhaka (Islam, 2016). These spaces have transformed into 'places' where especially, women and children can socialize with neighbours. Thus, social harmony has thrived in the Old Dhaka neighbourhoods from its foundation.

Despite this socio-cultural harmony, residents of Old Dhaka have long faced difficulties such as overcrowded living conditions, unsafe electricity connections, lack of clean drinking water, shrinking open spaces, and poor drainage systems due to rapid urbanization and population growth (Mukul, 2020). With the increasing needs and demands, new buildings are seen being erected often disregarding the rules and regulations. The future development of Old Dhaka is needed to be carefully planned after analysing the current state and extensive history it has. Therefore, RAJUK (Rajdhani Unnayan Kartripakkha-Capital Development Authority) has developed an urban redevelopment project for the Detailed Area Plan (DAP) 2022–2035 in a bid to resolve the existing civic problems in Old Dhaka. Seven possible places have been identified primarily at Islambagh, Chawkbazar, Moulvibazar, Bangshal, Hazaribagh, Kamrangirchar and Lalbagh for this urban redevelopment initiative (RAJUK, 2020). This chapter demonstrates the urban regeneration proposals of DAP 2022–2035 for two Old Dhaka neighbourhoods, i.e. Islambagh and Bongshal.

3 Methodology

3.1 Study Design

This research is based on Participatory Learning and Action (PLA) approach in the first phase and design demonstration in the second phase. As cited by Ericson (2006), people might have active participation in the planning phase by using the proper participatory approaches and developing realistic, contextual, and long-term interventions for heritage-led neighbourhood regeneration projects. Additionally, Li

Fig. 1 Map of Old Dhaka, indicating study areas (*Source* RAJUK (2010); modified by the authors)

et al. (2020) added that the design charrette or demonstration is usually a multi-day intensive planning process in which a group of professionals and users generate a comprehensive growth or development plan that represents the participation of a community that is engaged through several feedback loops. For this research, several participatory approaches were exercised, including focus group discussions, key informant interviews, transect walks and community mapping to identify the existing problems, their experience living with these challenges and their aspirations. Following that, a general guideline was developed based on their feedback and later, two urban design workshops were conducted with 50 participants to demonstrate the urban regeneration proposals of DAP 2022–2035 for two Old Dhaka neighbourhoods, i.e. Islambagh and Bongshal (see Fig. 1) among the identified seven neighbourhoods.

3.2 *Study Settings*

This research was conducted in two Old Dhaka neighbourhoods, i.e. Islambagh and Bongshal (see Fig. 2). Both of these neighbourhoods share similar socio-spatial features, history and heritage values as Old Dhaka, as well as the same civic challenges.

Islambagh Area: The study area in Islambagh is bordered by 'Water Works' Road to the north and east, by Islambagh Road to the south and by Posta Road to the west. The

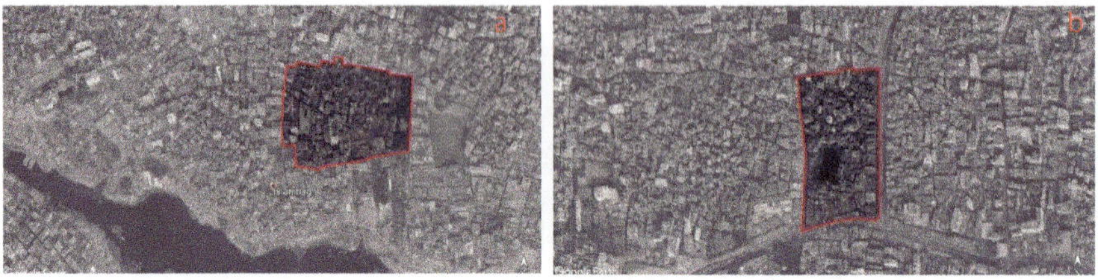

Fig. 2 Location map of **a** Islambagh and **b** Bangshal (*Source* Google Earth Image; modified by the authors)

entire area may be split into four primary groups based on how they are used, with built areas covering about 73.14%; roads, lanes, and pathways occupying 18.16%, ditches taking up 4.98% of the area, and the remaining 3.72% of the land being left undeveloped. In comparison to the rest of Dhaka, the region is exceedingly crowded, with weak economic and poor living conditions. They have access to extremely few and insufficient civic facilities. Islambagh's housing situation evokes images of congested brick buildings, some houses made of wood and bamboo, and a few makeshift sheds. The majority of the buildings are built in very close proximity. Ground floors are generally freed up for rentable storage, such as plastic materials storage. The connecting roads are not wider than 7 m (maximum). Usually, locals of all ages populate tea stalls, grocery shops, phone booths and even, bazaars around the mosques. Therefore, these spaces become more active hubs for social interaction.

Bangshal Area: The study area in Bangshal is bordered by Bangshal Road to the north and east, by Nawab Yusuf Road to the south and by Kazi Alauddin Road to the west. In this area, the built area is covering about 63.5%, roads, lanes, and pathways occupying 22.5%, water body (Bangshal Pond) taking up 8% of the area and the remaining 6% of the land is open and vacant land. This area has long been commercial, especially with wholesale commodities of all sorts. However, Bangshal Jame Mosque, a RAJUK enlisted heritage building, is located onsite. Residents in Bangshal confront the same issues as residents of other neighbourhoods of Old Dhaka, such as congested living conditions, unsafe electricity connections, shortage of clean drinking water, shrinking open spaces, and poor drainage systems. Narrow internal lanes, age-old residential buildings made up of bricks and shops and stores on the ground floor are typical features of this neighbourhood. Furthermore, the Bangshal Pond, which is not only historically significant but also a place for the community to socialize and relax, is a source of pride for the neighbourhood. Both locals and visitors come here to fish, occasionally for competition and other times simply for enjoyment. People having a relaxing plunge at this pond is a typical sight.

4 Results and Analysis

4.1 Challenges of the Old Dhaka Residents

Residents of Old Dhaka are dealing with a lot of challenges, according to early assessments with locals. During focus group discussions, a ranking of these challenges of Old Dhaka was prepared using the pairwise comparison method based on their severity and extent. The three leading problems of Old Dhaka, which include water logging due to a poor drainage system, congested access to buildings due to narrow lanes, and inadequate basic civic services due to unplanned infrastructure have been identified and examined thoroughly.

The findings revealed that Old Dhaka neighbourhoods have open, low-capacity drains (width 15–20 cm in several areas), through which sewage and stormwater are simultaneously carried away (see Fig. 3). Therefore, during the wet season, if it rains heavily for even 10–15 min, the surface will quickly become inundated. The overflowing drainage systems flooded the roads and streets frequently, and residents' regular lives became disturbed and interrupted. Additionally, it reduces the capacity of streets for vehicular traffic, resulting in obstructions to access and mobility.

Again, Land subdivisions and development projects in Old Dhaka have not been complying with Dhaka Building Construction Rules for a long time. Furthermore, this historic core is a good example of such an area where rapid population growth has led to dense development without enough availability of open space or even access roads. This causes access routes to become more congested and harder to enter buildings. It is extremely difficult even for two people to walk side by side on certain neighbourhood lanes away (see Fig. 4).

In this research, basic civic services were determined by engineering utilities; e.g. power supply, gas supply, water supply, etc., and community services; e.g. open spaces, schools, community centres, hospitals etc. The residents of Old Dhaka reported daily power outages and erratic water and gas service during the field study. In addition, the number of community services was insufficient in this area, resulting in frequent overcrowding. Moreover, neither the planning nor the location of these

Fig. 3 Drainage and water clogging as one of the major challenges in Old Dhaka

Fig. 4 Narrow lanes with poor accessibility as another major challenge in Old Dhaka

community services with one another was convenient. The shortages of parks, playgrounds, and public open spaces are prevalent throughout the neighbourhoods of Old Dhaka. Therefore, locals often hung out on rooftops, at building fronts, on street corners and at roadside tea stalls.

4.2 *Socio-spatial Qualities of Neighbourhoods in Old Dhaka*

The study identified several unique physical properties of the Old Dhaka Neighbourhoods, including the presence of minarets and bridges in Old Dhaka Cityscape, street-side platforms and stairs at building fronts, rooftops as a place of recreational activities, retail shops at the ground beside the narrow lanes (Fig. 5).

The superbly built minarets of Dhaka serve as landmarks and contribute to its popularity as "the city of mosques". Almost every street here has a minaret in its vista, providing not only clarity to users but also a memory aspect to the residents. In Old Dhaka, mosque premises were also identified as social gathering spaces. Old Dhaka also has bridges and connections between upper-level functions. Bridges and connections in the upper level enable social ties between neighbours and also recreate the street character at different levels to provide the essence of the ground.

Fig. 5 Minaret, connecting bridges, street-side platforms and retail shops in Old Dhaka

The streets of old Dhaka are always busy, even at night. The street vendors, shopkeepers, and hawkers keep these streets busy all day and night. Retail shops often serve a crucial role in fostering social interactions, preserving social capital, and acting as 'eyes on the street'. These retail outlets have to be incorporated at both the ground level and higher-level nodes and intersections.

The platform supported by a few steps, known as a 'roc' by locals, is a typical sight in front of Old Dhaka buildings, enabling passers-by to converse. The 'roc' may be recreated by using the same design elements on both the ground floor and the upper-level corridors, providing spaces for people to gathering, waiting, or just enjoying vibrant neighbourhood activities.

Like the streets, rooftops are the most used places in old Dhaka. Women spend most of their time on rooftops. Children play on the roofs, and adults use the rooftops to fly kites and pigeons. Due to the connectivity of the existing buildings here, there is an interconnectedness among the rooftops that enables social contact among neighbours in the upper ground. If the rooftops of the buildings can be kept connected, the current rooftop activity patterns can be maintained.

Moreover, throughout the study, heritage buildings, natural heritages, and historical pathways were identified in almost every Old Dhaka neighbourhood. These heritage buildings and natural heritages must be protected and preserved thoroughly during the redevelopment process. Additionally, those historical streets may be transformed into pedestrian-only pathways, prohibiting all vehicular access. Providing active social and communal activity spaces along these pathways can preserve these streets' present lively character.

4.3 Design Demonstration for Islambagh and Bangshal Area

During the design process, some suggestive design demonstrations were practised as an experiment to comprehend what is possible. Some of the design demonstrations are here:

Accommodating density: To accommodate the growing population and for the users' benefit, these areas need to be developed with higher density. According to the Floor Area Ratio (FAR) introduced by the Dhaka Building Construction Rules 2008, the building height has to be expanded to 15 storied to accommodate this higher density. Moreover, the Detailed Area Plan (DAP) 2022–2035 suggested that Block Development can figure out the appropriate number of residential units while allowing for open spaces and other public facilities. For design solutions, Block development can be a viable option by scaling down building heights at the periphery to blend with the context around the plots.

Widening roads and streets: As indicated before, peripheral and principal roadways must be widened to allow easy vehicular movement and accommodate emergency ambulances and fire engines. Keeping the periphery road for vehicular traffic, internal roadways can be converted to pedestrian-only with limited access for non-motorized vehicles. Food cart areas can be accommodated as social gathering spaces on the edges and corners of the streets and conserved for children, besides the open play areas.

Allowing existing commercial belt: Alongside the road lies a commercial belt (i.e. shops on the ground floors of every building), which is a major aspect of Old Dhaka. Instead of constructing a boundary wall and isolating the block area from its surroundings, the commercial belt and small manufacturers can preserve the existing fabric of the street of Old Dhaka. This will also help to integrate the new development with the existing ones rather than alienate them.

Allocating civic facilities: In addition to residential units, new types of programs as civic facilities can be recommended for the residential area based on the Dhaka Building Construction Rules, 2008 and DAP 2022–2035. With the limited amount of space, multiple-use spaces can be suggested to make optimum use and allow variety. There is the potential for overlap between co-working spaces, day-care facilities, and club facilities. Small areas on several levels can be developed for women to sit together and work on different homemade products. Spaces to sell various handcrafted goods can help the informal economy in these places.

Creating open spaces: Old Dhaka lacks open spaces, a prerequisite of any residential neighbourhood by the rules and regulations. Therefore, keeping open play fields and open spaces to allow for a variety of activities is crucial in the redevelopment process. These spaces need not be large green areas, instead, they can be different spaces, like streets on different levels or narrow streets that let all the functions happen while still keeping the existing fabric.

Preserving heritage and memory: In addition to designing the new buildings, it is important to keep the heritage and the memories of the existing scenario alive. Reconstructing the streets to include 'rocs', small spaces, verandas, and roofs at varying elevations, as mentioned before, can bring back memory spaces for the locals. The choice of how to link the heritage buildings or natural heritages (e.g. open space or water bodies) is also crucial. For example, in the Islambagh area, the Boubazaar can be conserved and additional events can be organized to help preserve the memory of these sites (see Fig. 6). Again, in Bangshal, the old Bangshal Jame Mosque and Historic Bangshal Pond can be connected by a pedestrian pathway with public spaces around the pond for both locals and visitors (see Fig. 7). Recreating minarets, arches and bridges can create memories for the people living in these areas.

5 Discussion

This research sought to establish a connection between two concerns pertaining to the historic city core: the desires of locals to make the neighbourhoods liveable by addressing the problems that have long plagued them, and the regeneration of the neighbourhoods while preserving their unique social and spatial characteristics.

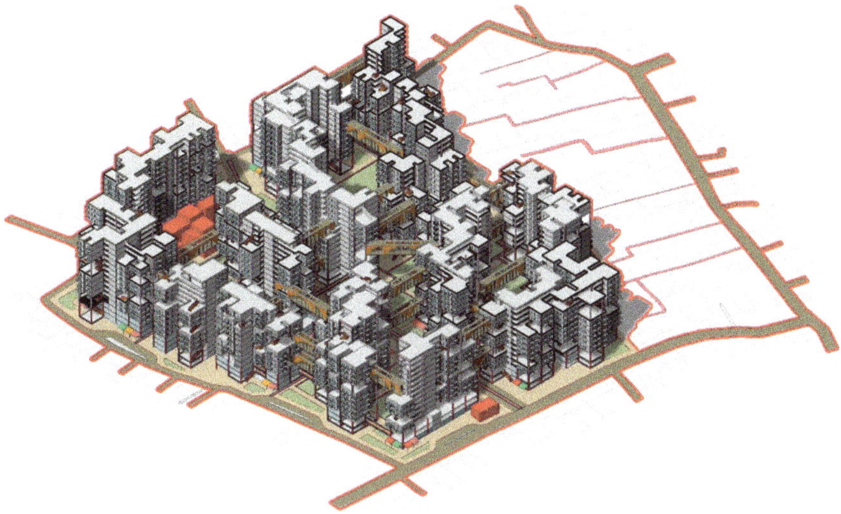

Fig. 6 Block development design option for Islambagh

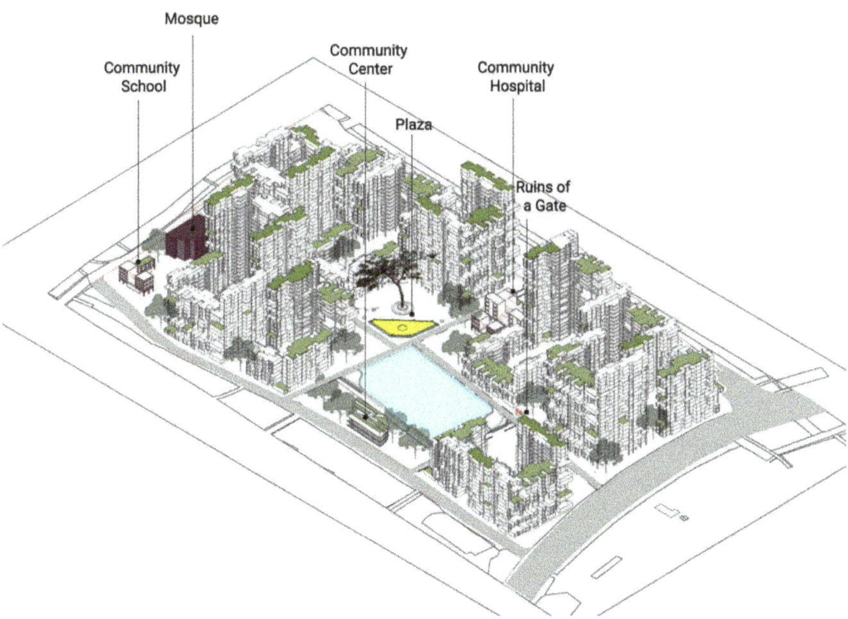

Fig. 7 Block development design option for Bangshal

For this, before beginning any revitalization efforts in the neighbourhoods of Old Dhaka, this research used a participatory method to determine local potentials and challenges.

Islambagh and Bangshal residents have complained mostly about Poor drainage, narrow lanes and unplanned infrastructures. Block development was proposed as a means of ensuring that each neighbourhood had sufficient access to public services. With this design option of block development, the potential for creating additional green spaces was identified, therefore resolving the drainage overflow issue. Rahman (2020) illustrated the advantages of green areas in terms of water retention and percolation, advising landowners to develop multiple plots together to maximize the amount of open space on the ground in Dhaka. Last of all, widening the peripheral and principal roadways were recommended to solve the problem of narrow access routes. To illustrate the redevelopment process for the locals, all of these interventions, along with other socio-spatial elements, were integrated into the design exercises. Correspondingly, in Shenjing village in Guangzhou, architectural visualization and drawings were prepared to take feedback from the community people regarding the redevelopment project (Li et al., 2020).

As cited by, Tokey et al. (2020), even though well-intentioned, not all instances of redevelopment projects or regeneration activities end up being beneficial to the local community. Consequently, urban planners need to emphasize the relevance of learning more about how individuals participate in the development of their communities and work to conserve their history and heritage (Tweed & Sutherland, 2007). Unfortunately, the majority of urban development initiatives in Dhaka have been carried out without considering the local context of the neighbourhoods (Daisy et al., 2013). Therefore, this study was conducted to ensure that the participation of local people, particularly those of Islambagh and Bangshal, in the redevelopment of Old Dhaka would be given due consideration.

6 Conclusions

To conduct an urban regeneration project in a historic core, local public participation may serve as a driving factor in resolving disputes amongst the stakeholders. This public participation can be easily found in the form of collective workshops in China (Li et al., 2020), 'Machizukuri' in Japan (Okyere et al., 2019) and citizen-led planning in other countries (Rock, 2018) for community development and redevelopment activities. Through public participation in neighbourhood regeneration projects, a middle ground should be maintained between the traditional top-down approach and the more organic bottom-up approach. Therefore, in the case of redevelopment planning for a historic core like Old Dhaka, the priorities and values of local people should be considered. Subsequently, Old Dhaka will be redeveloped into a more liveable shape as a direct and indirect outcome of local people's values for the benefit of its neighbourhoods. When planning to redevelop a dense urban area like Old Dhaka, it is also important to consider all of the essential public amenities critically and comprehensively. With this in mind, the locals need to be actively involved in any plans for regenerating the neighbourhoods of Old Dhaka to ensure that its history, heritage, culture and traditions are preserved properly without compromising all the liveability indices for them.

Acknowledgements The authors gratefully thank all of the respondents from two Old Dhaka neighbourhoods for their kind assistance during the research. The students of the Department of Architecture from the University of Asia Pacific (UAP) are also acknowledged by the authors for their support of the survey and design workshop.

References

Ahmed, N. (2021). Dhaka City's unbridled expansion: Which direction is it going to grow in the future? *The Daily Star.*

Akter, T., Rashid, M., Mamun, M. M., Sadmani, I., Akther, S., Islam, I., & Hossain, M. D. (2017). Prospect of redevelopment of the Showarighat Area of Dhaka Central Jail. *Journal of Bangladesh Institute of Planners, 10*, 25–40.

Ashraf, K. K. (2022). DAP 2022–2035: Rajuk's plan to fix Dhaka might just work. *The Daily Star*.

Badland, H., Whitzman, C., Lowe, M., Davern, M., Aye, L., Butterworth, I., & Giles-Corti, B. (2014). Urban liveability: Emerging lessons from Australia for exploring the potential for indicators to measure the social determinants of health. *Social Science & Medicine, 111*, 64–73.

Chen, Y., & Rosenthal, S. S. (2008). Local amenities and life-cycle migration: Do people move for jobs or fun? *Journal of Urban Economics, 64*(3), 519–537.

Choi, Y. S., & Seo, B. K. (2020). Understanding contextualised liveability from the bottom up: A qualitative analysis of the participatory planning proposals in Daegu, South Korea. *Architectural Research, 22*(3), 75–84.

Clark, J., & Wise, N. (2018). *Urban renewal, community and participation.* Springer.

Daisy, N. S., Debnath, A. K., Rahman, N., Islam, I., & Hasan, M. U. (2013). Community perception on the redevelopment of Old Dhaka based on local solutions. In S. Jahan & A. K. M. A. Kalam (Eds.), *Dhaka metropolitan development area and its planning: Problems, issues and policies.* Bangladesh Institute of Planners (BIP).

Dave, S. (2010). High urban densities in developing countries: A sustainable solution? *Built Environment, 36*(1), 9–27.

Ericson, J. A. (2006). A participatory approach to conservation in the Calakmul biosphere reserve, Campeche, Mexico. *Landscape and Urban Planning, 74*(3–4), 242–266.

Fatemi, N., & Islam, N. (2014). *Learning from precedents: Emerging urban planning practices towards livable communities in Dhaka.* Paper presented at the 2nd International Urban Design Conference on Cities, People and Places (ICCPP-2014), Colombo, Sri Lanka.

Glaeser, E. (2012). *Triumph of the city: How our greatest invention makes us richer, smarter, greener, healthier, and happier*. Penguin Books.

Hankins, K. B., & Powers, E. M. (2009). The disappearance of the state from "livable" urban spaces. *Antipode, 41*(5), 845–866.

Hee, L. (2015). Can dense cities really be liveable? In S. T. Rassia & P. M. Pardalos (Eds.), *Future city architecture for optimal living* (pp. 81–104). Springer.

Iffat, W., & Fatemi, M. N. (2012). *Adaptive roof tops as a multifunctional breathing space: The Old Dhaka context.* Paper Presented at the International Seminar on Livable Space (IsLivas)—2012: Creating Space for Better Life, Jakarta, Indonesia.

Islam, Z. H. (2016). Changing housing scenario of Dhaka and its impact on social interaction. *Cities People Places: An International Journal on Urban Environments, 1*(2), 16–29.

Islam, N., & Fatemi, M. N. (2015). *Spaces for social cohesion in planned city: A search for inspiration in indigenous and new spontaneously developed settlements of Dhaka.* Paper Presented at the 3rd International Conference on Cities, People and Places-ICCPP-2015, Colombo, Sri Lanka.

Leby, J. L., & Hashim, A. H. (2010). Liveability dimensions and attributes: Their relative importance in the eyes of neighbourhood residents. *Journal of Construction in Developing Countries, 15*(1), 67–91.

Lee, K. Y. (2021). Factors influencing urban livability in Seoul, Korea: Urban environmental satisfaction and neighborhood relations. *Social Sciences, 10*(4), 138.

Leh, O. L. H., Aziz, M. H. A., Mahbot, N. M., & Marzukhi, M. A. (2020). A study of urban liveability in a city and a suburban—Case study: Kuala Lumpur and Puncak Alam, Malaysia. *Journal of Surveying, Construction and Property, 11*(2), 16–26. https://doi.org/10.22452/jscp.sp2020no1.2

Li, X., Zhang, F., Hui, E. C.-M., & Lang, W. (2020). Collaborative workshop and community participation: A new approach to urban regeneration in China. *Cities, 102*, 102743.

Looker, B. (2012). Visions of autonomy: The new left and the neighborhood government movement of the 1970s. *Journal of Urban History, 38*(3), 577–598.

McGee, T. (2010). Building liveable cities in Asia in the twenty-first century research and policy challenges for the urban future of Asia. *Malaysian Journal of Environmental Management, 11*(1), 14–28.

Mowla, Q. A. (2004). Memory association in place making: Understanding an urban space. *Protibesh, 9*, 52–54.

Mukul, M. (2020). Old Dhaka to be changed. *The Daily Industry.*

Nahrin, K. (2018). Urban development policies for the provision of utility infrastructure: A case study of Dhaka, Bangladesh. *Utilities Policy, 54*, 107–114.

Oh, Y. K., Choi, H. S., & Lee, E. J. (2013). Changes in urban policy paradigm and composition of community indicator. *The Korean Association for Governance Studies, 23*(3), 1–21.

Okyere, S. A., Diko, S. K., Abunyewah, M., & Kita, M. (2019). Toward citizen-led planning for climate change adaptation in urban Ghana: Hints from Japanese 'Machizukuri' activities. In P. B. Cobbinah & M. Addaney (Eds.), *The geography of climate change adaptation in urban Africa* (pp. 391–419). Springer International Publishing.

Rahman, T. (2020). *A study on the potentials of green blue infrastructure (GBI) network for Dhaka City in Bangladesh.* Master thesis, Osaka University, Osaka, Japan.

Rahman, T., & Fatemi, M. N. (2021). Liveability dimensions in new town developments: An overview of Senri New Town and Purbachal New Town. *Journal of Contemporary Urban Affairs, 5*(2), 221–233. https://doi.org/10.25034/ijcua.2021.v5n2-6

RAJUK. (2010). *Detailed area plan (DAP).* Rajdhani Unnayan Kartripakkha (RAJUK).

RAJUK. (2020). *Urban redevelopment in Old Dhaka: Project synopsis.* Retrieved from Dhaka, Bangladesh.

Rock, A. E. (2018). Citizen participation and public funding in Ohio. In J. Clark & N. Wise (Eds.), *Urban renewal, community and participation: Theory, policy and practice* (pp. 181–197). Springer International Publishing.

Said, S., Zubir, S. S., & Rahmat, M. (2014). Measuring physical changes in an urban regeneration scheme. In N. Marchettini, C. A. Brebbia, R. Pulselli, & S. Bastianoni (Eds.), *The sustainable city IX: Urban regeneration and sustainability* (Vol. 191, pp. 1165–1174). WIT Press.

Salaripour, A., Mehrjou, M., & Jalilisadrabad, S. (2021). Evaluation of urban neighborhood regeneration based on urban liveability criteria case study: Kolapa neighborhood, Hamedan City. *International Journal of Architectural Engineering & Urban Planning, 32*(2). https://doi.org/10.22068/ijaup.573

SRD. (2022). *Cities with the highest population density worldwide in 2022.* From Statista Research Department (SRD).

Stafford, L., & Baldwin, C. (2018). Planning walkable neighborhoods: Are we overlooking diversity in abilities and ages? *Journal of Planning Literature, 33*(1), 17–30.

Throsby, D. (2005). *Cultural heritage as financial asset in strategies for urban development and poverty alleviation.* Paper presented at the International Conference for Integrating Urban Knowledge and Practice, Gothenburg, Sweden.

Tokey, A. I., Shioma, S. A., Muniruzzaman, M., Munir, D. H., Akther, M. S., & Islam, I. (2020). Redevelopment of a dense area: A participatory planning approach for regeneration in Old Dhaka, Bangladesh. *Journal of Regional and City Planning, 31*(3), 217–236.

Tweed, C., & Sutherland, M. (2007). Built cultural heritage and sustainable urban development. *Landscape and Urban Planning, 83*(1), 62–69.

Vine, D. (2012). *The neglected dimension of community liveability: Impact on social connectedness and active ageing in higher density accommodation.* Doctoral thesis, Queensland University of Technology.

Woolcock, G. (2009). *Measuring up?: Assessing the liveability of Australian cities.* Paper presented at the National Conference on the State of Australian Cities, Perth, Australia.

Chapter 7
The New Strategy of Urban Village Regeneration: The Comprehensive Improvement Project in Jingle Village, Shenzhen, China

Pengyu Chen and Tim Heath

Abstract The urban villages are a common yet heterogeneous informal living space in China's developed cities, characterised by high density, a high degree of autonomy, and difficulty in being regulated by the government. As a provider of low-cost residential accommodation, they have contributed significantly to the successful growth of cities, however, they have also posed a number of problems such as hygiene and security. During the period of rapid industrialisation and urbanisation in China, the government has often considered demolition and redevelopment to be a convenient and cost-effective way of regenerating the urban villages. However, with the recent shortage in the supply of land resources and soaring land prices, the demolition and redevelopment approach has created a series of social and policy issues. Recently, comprehensive improvement—a way of beautifying the built environment and optimising management without altering the building construction—has been introduced as an alternative approach to regenerate urban villages and it is anticipated as a sustainable model of urban regeneration. The planning and implementation process involves multiple stakeholders, including the government, developers, villagers, tenants, etc. These different actors, however, all have different interests and expectations in the processes of planning, implementation, and operation. While contributing to the sustainability of the regeneration projects, many potential problems have also arisen, some of which have led to small-scale open protests. This chapter uses the case of Jingle Village in Shenzhen as a case study to examine the interests of different stakeholders at different stages of the comprehensive improvement process and the problems faced in the implementation of the project. The chapter critically assesses and analyses the positions and roles of the various stakeholders at different levels based on in-depth semi-structured interviews and resident questionnaires, in order to assesses the sustainability of the new regeneration model.

P. Chen (✉)
Department of Architecture & Urban Planning, Shenzhen University, Shenzhen, China
e-mail: pengyuchen@szu.edu.cn

T. Heath
Department of Architecture & Built Environment, University of Nottingham, Nottingham, UK

E. Mangi et al. (eds.), *Innovative Public Participation Practices for Sustainable Urban Regeneration*, Urban Sustainability, https://doi.org/10.1007/978-981-99-9595-0_7

The findings can contribute significantly to the sustainability of urban regeneration strategies in China and even the global South. The analysis of the different levels of stakeholders also will be of great value in informing many different types of regeneration policies.

1 Introduction

With the decrease of the 'rough and tumble' economic model in China, the attitude of the state and first-tier cities towards land resources has also changed from vigorous expansion to land storage redevelopment. The lack of available development land led to the demolition and rebuilding of old buildings or the renovation of old buildings becoming the main redevelopment mode in the city (Chiu-Shee & Zheng, 2019). In the early of twenty-first century (2005–2018), a large number of informal settlements like urban villages have been demolished and rebuilt in China as a mainstream method of its urban renewal plan (Pan & Du, 2021). To date, there has been a lot of research on the redevelopment of informal settlements at both the global scale (Doshi, 2013; Ren, 2017) and in China (Liu et al., 2016; Wu et al., 2012). However, there is not much research on the new comprehensive improvement projects, especially on the role of the stakeholders and their contributions and conflicts.

There is also significant research on the urban villages in China, most of them focusing on their influence cities and their physical conditions. Indeed, Wu (2016) undertook detailed research on the dwellers' conditions and the tenancy informality of Chinese urban villages. Tan et al. (2019) also evaluated residents satisfaction with market-oriented urban village transformation, whilst Gao et al. (2020) argued that the type of street within the urban village has considerable influence on accessibility, functional mix, and the urban interface.

Since comprehensive improvement is a new policy to Shenzhen that has not yet produced large-scale impacts and contradictions, there is almost no research on the detailed process of comprehensive improvement or the characteristics of all of the parties involved. As a new urban renewal policy, it is necessary to thoroughly study its specific process and problems, especially the positions of various stakeholders as the planners and implementers of its implementation, so as to promote the sustainable development of the policy.

In similar studies, Jiang et al. (2020) discussed the interest conflict of stakeholders in demolition and reconstruction type of renewal, and Hin and Xin (2011) discussed the role and influence of local governments and their negotiation methods in urban village renewal. Li's research (2021) placed the study of a comprehensive improvement project in the context of state behaviour, and suggested that the project was a way for state actors to use informal operations to strengthen their perceived power security and to improve the spatial quality of informal property, so as to achieve government goals. However, none of this research focussed on these projects from the perspective of macro policy. Therefore, at the level of urban renewal, there are many articles that study stakeholders from the perspective of the government or the residents, but very few articles that discuss their participation modes from the perspective of developers. Against this background, this chapter intends to bridge that gap, by disentangling people's misunderstanding of the influence and connotation of the developer in the implementation of comprehensive improvement.

The research presented in this chapter suggest that, although the developers are profit-oriented, they can also make a considerable contribution to the sustainability of the city in the context of the social policy of comprehensive improvement and have a positive impact on the long-term interests of all parties in the city, such as through their rental management mode. This chapter also adds to the thinking on the governance of informal settlements by revealing the role and influence of stakeholders at different stages of the process, by conducting in-depth interviews and questionnaires with multiple stakeholders involved in the implementation process. This has enabled a better understanding of potential issues that arise during the negotiation and implementation phases. In addition, the empirical evidence provided supports the sustainable urban development policies of developing countries and local governments in the global South. The remainder of this chapter addresses several issues, including a literature review of urban villages in China, the changes in urban village policy and the provisions of the new policy for sustainable urban development. In addition, the empirical methodology of the study is presented as well as the implementation approach and the research results and findings together with the limitations of the study.

2 Informality of the Urban Village and Policy Changes

Informal settlements are a mode of production that globally accommodates more than a billion people (Dovey et al., 2020). Indeed, the concept of informal settlements is a spontaneous reaction to the failure of national laws and regulations to meet the basic needs of poor people (Dowall, 1991; Fekade, 2000). Significantly, Davis (2006) identifies how because lack of regulation they are often seen as a unsanitary places with accumulated poverty. Informal housing is, however, often also a foothold for newcomers to the city because of the lower costs of living, and the cohesiveness of immigrant populations (Cybriwsky, 2013; Dupont et al., 2015; Huchzermeyer, 2011).

In 2003, Zhang et al. (2003) studied the issues of urban villages from the perspective of self-help housing. However, due to the uniqueness of property rights and the operation of Chinese urban villages, it is hard to reconcile them within the framework of theories of informal housing or self-help housing (Wu, 2016). Most urban villages in China are a foothold for migrant workers and Wang et al. (2009) suggest that they can significantly influence many development indicators of a city. An 'urban village' in the Chinese context refers to those traditional villages which have lost their farmland (usually purchased compulsorily by the government), in a geographic location that has gradually been surrounded by urban development. The subsequent dichotomy, of rural and urban, thereby creating two opposing statuses of space and inhabitation. The economy of rural areas relies on its production of land, which mainly refers to agriculture and animal husbandry, however, secondary and tertiary industries are the main financial support of urban areas. Due to the high cost of compensation and fees for relocating the farmers, they have often been left 'enclosed' by urbanisation as an autonomous zone within an expanding city and outside the control of the local government. Because of the high autonomy of the villager and the high value of land in modern city, the villagers have often tried to build apartment buildings as high as possible on their own land for the rental income, since they no longer had income from farmland. Significantly, the high density of buildings in the urban villages can create considerable income for the villagers but also cause a series problem, such as poor living environments, poor levels of hygiene, poor construction quality, and potential safety hazard from fire and criminals (see Fig. 1) (Wang et al., 2010). However, beyond these obvious factors, social conflict, and the obstacle that urban villages cause to a city's further urban development are serious problems facing many cities.

Due to these various problems within the urban villages, Shenzhen government issued their 'demolish and rebuild' policy in 2004 (see Fig. 2). While this policy intended to eliminate urban villages and to accelerate urban construction it actually resulted in a large number of new problems emerging. These included the exacerbation of gentrification, the loss of accommodation for the city's cheap labour force, and the decline of urban attractions of Shenzhen. The subsequent new 'comprehensive improvement strategy' launched in 2018 has attempted to establish new solutions to these previous problems.

Fig. 1 Typical physical context of an urban village in Shenzhen (*Source* Photo taken by the authors)

Fig. 2 The demolition of Caiwuwei South Village urban renewal project in Shenzhen under the 'demolish and rebuild' strategy (*Source* Photo taken by the authors)

As China's central government adjusted its social direction, the Sixth Plenary Session of 16th CPC Central Committee adopted a 'decision on several major issues concerning the construction of a harmonious socialist society' in October 2006 (Xinhua News Agency, 2006). Subsequently, the crude economic development model of exchanging nature for economic growth was no longer advocated, and the phenomenon of large scale 'demolition and rebuild' began to decrease. The main contradiction in Shenzhen at that time was not only the shortage of development land but also how to guarantee the availability of a sustainable low-cost labour force so as to maintain the city's rapid economic development. This went against the previous strategy for urban villages of 'demolish and rebuild' and therefore, the new policy followed the trend. As a result, in March 2019, Shenzhen Municipal Bureau of Planning and Natural Resources officially released the Overall Planning for Comprehensive Improvement of Urban Villages (old villages) in Shenzhen 2019–2025 (hereinafter referred to as the Overall Plan of Urban Villages) (Shenzhen Municipal Bureau of Planning & Natural Resources, 2018). The Shenzhen government defined the Comprehensive Improvement Projects (CIP) as: "mainly including improvement of fire protection facilities, infrastructure and public service facilities, street facades, environmental renovation and energy-saving renovation of existing buildings, but do not changing the main structure and use function of the buildings" (Shenzhen Municipal Bureau of Planning & Natural Resources, 2019).

The new CIP policy solved two main problems, firstly, the poor physical conditions of the urban villages, and secondly the negative impact of the previous D+R strategy. In terms of the problems of the urban villages themselves, the Overall Plan of Urban Villages specified in article 6 that one of its key aims was to "*eliminate safety hazards in urban villages and improve the living environment and supporting services.*" In addition to slowing down the process of gentrification, the Shenzhen government also aimed to convert the urban villages into a reserve of affordable housing for the immigrant workers who were making a significant contribution to the rapid development of the city. Meanwhile, in order to prevent the eviction of tenants due to increasing rents after comprehensive improvement, the Overall Plan of Urban Villages also stated that relevant government departments should strengthen the lease management of residential accommodation in the urban villages. The continued increase of rents was however inevitable as it is related to the cost of unified rent and operational costs of the developers. The currently completed comprehensive improvement projects include the youth apartment project in Jingle Village led by the developer Vanke and both the advantages and disadvantages of this approach will be discussed in the following sections.

3 Urban Village Policy of Shenzhen

3.1 Shenzhen's Strategy

The 'Shenzhen Urban Renewal Measures' were promulgated in 2009, immediately following the implementation of the 'Three Old Urban Renewal Plans' in Guangzhou in 2008. Shenzhen's current approach to urban renewal can, however, be divided into three main categories: demolition and rebuild; functional changes; and comprehensive redevelopment. The whole strategy is based on the 'Urban Renewal Measures' at the core, using a combination of government guidance and market operation to renew the areas. The functional changes are mainly about changing the use of the old factory buildings or the factory buildings in the villages, such as changing them into creative parks or innovative office areas like the OCT project (A creative industrial park project transformed from old industrial buildings). Significantly, demolition and rebuild and comprehensive redevelopment have mainly been applied to the renewal of the city's urban villages. As a result, the renewal project of urban villages in Shenzhen can roughly divided into two stages, the first is the large-scale demolition and reconstruction that began in 2004, and the second is the comprehensive improvement that began in 2018. The 'demolish and rebuild' (D+R) strategy was put forward under the background of both a lack of constructible land and urban village issues (poor living conditions, high crime rates, etc.) (Yangteng, 2004; Zhou, 2004).

D+R was the mainstream direction of urban renewal until 2018, with the advantages of quick profit, positive market, and a mature market. As a result, there are many completed projects that have achieved good results in terms of meeting the new functional needs of Shenzhen city as well as financial returns for the government. This approach aimed to unify Shenzhen's urban context and completely eliminate the countryside, however, such a utopian fantasy was not realized, on the contrary it caused many new problems such as the displacement of tenants and exacerbated gentrification (Liu et al., 2018; Uehara, 2008). Consequently, Shenzhen's government issued its new urban village policy in 2018 which introduced the new CIP strategy for the urban villages, to meet the housing demands of low-income groups in order to maintain the sustainable development of the city.

Shenzhen has many years of experience in urban renewal, and the whole regeneration system and policy framework is well established. Although many projects are driven by market, the government's control is still very strict. From macro-level regulations, management, to specific operations and techniques, the entire process is regulated by relevant regulations and guidelines. There is also a dedicated Renewal Bureau under the Shenzhen Planning and Land Commission, which is responsible for the operation of specific regeneration projects. The Bureau is responsible for the implementation of urban renewal projects, not only with the district government, but also with the municipal government. In contrast to Guangzhou, where the implementation of regeneration projects is mostly handled by the district government, the Shenzhen municipal government has a much stronger control over regeneration (see Fig. 3).

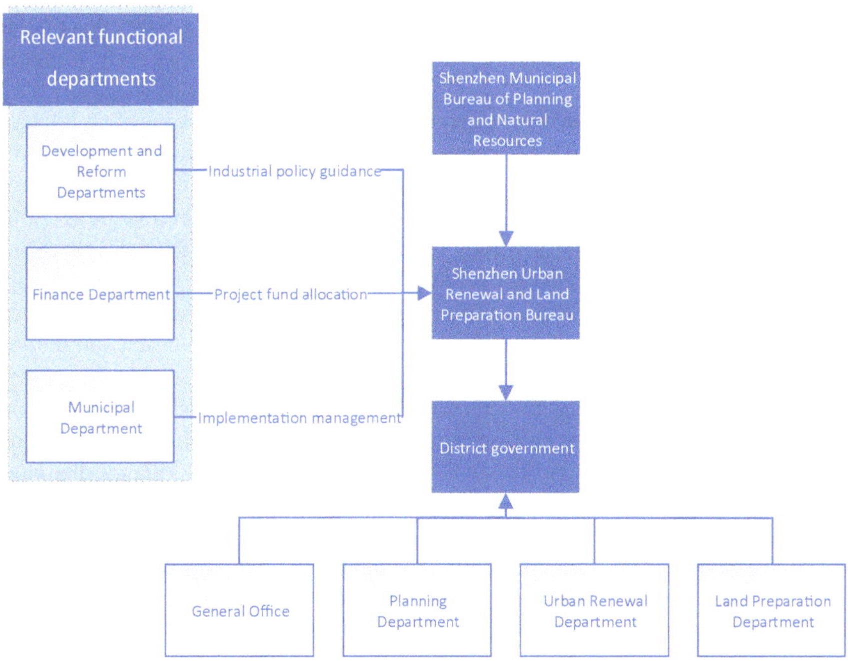

Fig. 3 Structure of the urban renewal department within Shenzhen municipal government (*Drawn* By the authors)

3.2 Different Model of CIP in Shenzhen

There are three main models for urban renewal policies and strategies: resident self-financing; government overall planning; and public–private partnerships (Wang et al., 2021). In terms of Shenzhen, according to interviews with government officers and a review of government documents, the current unified rent model in Shenzhen can be broadly divided into three modes: Unified Lease and Management Mode of Enterprises (UME); Unified Lease and Management Mode of Government (UMG); and Village self-governance (UMV). There are many kinds of unified lease models in Shenzhen's urban villages, among which the representatives of the UME mode are developer Vanke's Nantou Ancient Town and Jingle Village. These three modes can be defined as follows:

1. The UME mode is a typical model of enterprises or private leasing, which refer to enhancing the overall physical environment by renovating the interior and exterior of an existing urban village building without changing the main structure. The upgraded housing can be made profitable by increasing the rent. From a participation point of view, the UME mode involves the building acquisition, construction to the final operation, and management being undertaken by a company or organization who are also responsible for the main investment and benefit from the rental income.
2. The UMG mode also involves the refurbishment of buildings in urban villages both inside and out, the difference is that the apartments are primarily for government-designated people, and their main function is to be used as flats for talented people, reducing the cost of renting for them. The UMG mode is dominated by the government or wholly state-owned enterprises, that take charge of the construction and operation of building acquisition. Due to the complexity of the market and the difficulty of policy implementation, there are also urban villages in which the government and enterprises cooperate to carry out unified leasing.
3. UMV is simplest method to achieve the goal of comprehensive improvement. The investment and construction are done by the village itself, rarely involving other companies or the government. Most of this is done by villages that have the strong economic foundation to do so, on the one hand they are free to choose what they want to improve and on the other hand it is for the beautification of the village environment. This is beneficial to their own livelihoods and rental income and keep the administration authority themself (Table 1; Fig. 4).

Table 1 Three modes of urban renewal in Shenzhen (by the authors)

	Investor	Management	Rent profit	Preparation	Construction
UME	Enterprises	Enterprises	Enterprises	Government	Enterprises
UMG	Government	Government	Government	Government	Enterprises
UMV	Village	Village	Village	Village	Village

Fig. 4 Jingle Village urban renewal project in Shenzhen before and after under 'UME' type of comprehensive improvement strategy (*Source* Photo taken by the authors)

4 Analytical Framework

4.1 Research Method

This chapter focuses on Jingle Village in Shenzhen, as a representative of a comprehensive improvement project, to examine the interests of different stakeholders at different stages of the comprehensive improvement process and the problems faced in the implementation of the project. The original research presented in this paper was conducted from July to December 2020 as part of a larger study of sustainable redevelopment of Chinese urban villages. We conducted several in-depth interviews, including 2 developers, 2 government offices, 2 designers, and several villagers and residents. Each of these had been involved in the comprehensive improvement project with each interview lasting from 30 min to 2 h. This chapter critically assesses and analyses the positions and roles of the stakeholders at various levels through these interviews and questionnaires with residents, in order to assess their contributions to and conflicts between them in the renewal project. The results of this study will contribute significantly to the sustainability of urban renewal strategies in China and the global South. The analysis of the different levels of stakeholders will also be of great value to many different types of renewal policies.

4.2 Identification of Stakeholders

This study applies a template from the design consultancy field to make an initial classification of stakeholders: developers; designers; inhabitants; and policy makers.

High Influence—High Interest: Developers

In most cases, the developers are the investors in the urban renewal projects, and they significantly affect the landscape, urban space, and the supply and design of family space (Kriese & Scholz, 2011). As the main promoter, implementer and beneficiary of the policy, the developer plays a decisive role in CIP projects, both in terms of influence and relevance. In a sustainability analysis, it is important to apply a 'close attention' strategy to this type of stakeholder and also to consider how well the policy is accepted by these stakeholders.

High Influence—Low Interest: Government

The government plays the most important role in the governance structure of urban renewal, which directly affects the respective planning strategy (Zheng et al., 2014). Although governments are policy makers, they do not generally exert their influence as long as their basic interests are not compromised. Therefore, for these groups, they can either change the regulatory system to make it more relevant and transform themselves into key stakeholders or accommodate their interests within the framework of the current regulatory system.

Low Influence—High Interest: Residents

This includes the affected site residents, including visitors, who may not have a real impact on the implementation of the policy (in the context of city-wide urban renewal), but may sometimes act aggressively to protect their own interests. As the final customers of the community after urban renewal, the residents are the key stakeholders. Whether the urban renewal of the community is sustainable or not will keep affecting their daily life, meanwhile their preference and behaviours can have great impact on the decision-maker (Zheng et al., 2014). In addition, they contribute significantly to the sustainable development of the community (Bromley et al., 2016), which is why it is important to give special attention to these groups in terms of information disclosure and accessibility.

Low Influence—Low Interest: Designers

Designers are a low-influence, low-interest group in the implementation of policy, but as a design consultancy group they have a deep understanding of policy, and they are deeply involved in the process of policy implementation. Their understanding of the project can complement the gap between the developer and the residents, receiving the developer's intentions from the designer's point of view, while considering the lives of the users. This group's opinion is therefore a more objective complement to the developer's intentions (Fig. 5).

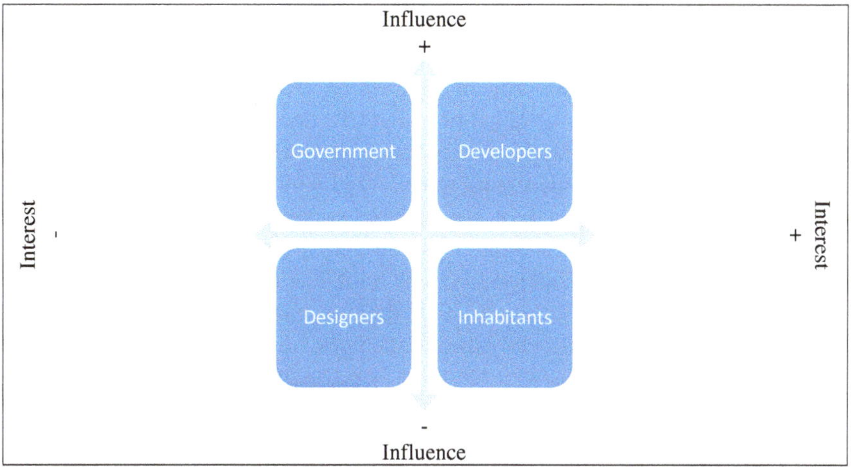

Fig. 5 Stakeholders of comprehensive improvement (*Drawn* By the authors)

5 The Process of Implementing Comprehensive Improvement: The Role of the Government and the Developer

For the new concept of 'comprehensive improvement', the guidance and detail in the Overall Plan of Urban Villages is too simple and general compared with the actual process and difficulty of implementation. In this section, the new policy will be described in terms of its process and implementation following a number of interviews with key developers. Some objective facts and problems which are not reflected in the government documents will also be revealed.

The chapter begins with a compilation of interviews and reviews from various sources to provide as complete and objective a description as possible of the entire comprehensive improvement process. This is followed by an analysis of the problems encountered in the process in the light of the particular circumstances mentioned in the interviews, with a large number of quotations from the interviews used to maintain objectivity.

5.1 Overall Implementation Process and Issues

The government, developers, and the village committees are the main players in the whole comprehensive improvement process. This tripartite relationship involves the government as the investor, the developers as agents of construction, and the village committees as the owners. The whole process of CIP can be divided into four stages: planning; negotiation; implementation; and operation. The comprehensive improvement process tends to be initiated in one of two ways, either the government actively seeking developers to involve them in the project, or the developers taking the initiative to undertake the project due to the potential of the long-term rental apartment market. It should be noted here that CIP does not necessarily mean that the urban villages will contain long-term rental apartments, it may just change the infrastructure, and the specific management could still be undertaken by the individual villagers themselves. In the negotiation stages of each project, the government, the developer, and the village committee conduct complex negotiations to determine a series of issues such as compensation, time limit, and the design scheme.

Close inspection of the results of CIP suggests that there are three potential outcomes (see Fig. 6). Firstly, only the comprehensive improvement of the infrastructure in the urban villages with the tasks being assigned to the district government by the municipal government. Secondly and potentially the most ideal result, where developers have completed the improvements and taken over the buildings as landlords. This has improved the surrounding environment and optimized the management of the rental market. Thirdly, in some villages, which have not been included in the CIP or the plan has not yet commenced, some developers have seized the opportunity to generate considerable profits by turning existing buildings into long-term rental apartments. Here, outside of any government control, developers have negotiated with the village committee directly and taken over the operation of the buildings after systematic rehabilitation and redecoration. The whole process of comprehensive can be divided into three stages: planning; implementation; and operation and these will be discussed in more detail in the following sections.

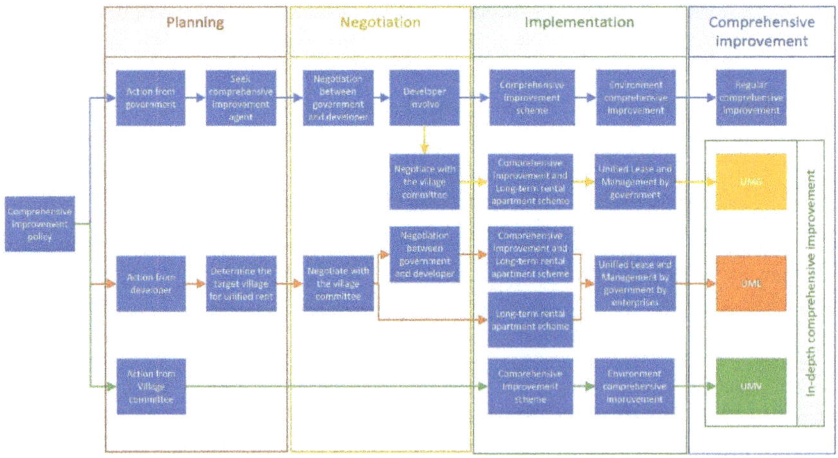

Fig. 6 The process of implementing comprehensive improvement (*Drawn* By the authors)

5.1.1 Planning

According to interviews with the government, there are three types of CIP, the first of which is the improvement of the village itself (UMV), which is organised by a strong village joint-stock company and managed by the village itself after the CIP. The government official interviewed said that: "*This is one of the more common methods at the moment. This approach may be more fragmented.*"; The second is Unified Lease and Management Mode of Government (UMG), as a supplement to the social housing system, through the unified management and operation of the housing system by government; the third type is Unified Lease and Management Mode of Enterprises (UME), the official said that: "*Because of the gradation of the rental market, different income groups will choose different urban villages to live in. Through corporate operation, it may be more upscale in terms of decoration and management, and the rental income of the company will be relatively higher.*" Another official said that they would not direct all urban villages to become long-term rental flats, but would leave the change to the market. These information from interviews with government officials basically confirms the three types of CIP in the previous section of the review.

For a developer, there are two ways that they can be involved in the process of comprehensive improvement (see Table 2): '**Zero Agent Construction**' (an agent construction project with zero agent-construction cost); and '**Open Bidding**.' In the 'zero agent construction' model, where the developer is actively willing to take on the task of construction as part of comprehensive improvement, the government is able to choose the developer directly from the government's supply list of developers. In this way, the main motivation for developers is to benefit from the long-term rental apartment market after the works are complete. In this process, the developer cooperates with the government and undertakes the improvement works as an investment with the financial return coming through the future rental returns from managing the property. The second model of 'open bidding' is higher risk for the developers with potentially much lower returns, because the process of CIP may not actually be profitable due to the stringent requirements and controls that the government places on the developer as seen in Article 22 of the document which states that: "...request the district government combined the original rural collective economic organizations or encouraging large enterprises, which have socially responsible and enough strength, not for the purpose of pursuing profit, actively participate in the comprehensive improvement of urban villages".

The government has no direct control over the developers, however, due to the government's rights to assign other projects in the city and the monopoly of information, it is actually difficult for developers to directly refuse the invitation of some government projects, even if these projects are unprofitable municipal projects. As such, the role of comprehensive improvement construction agent may not be lucrative for developers. one of the developers said: "*...usually we do CIP projects in order to have a good relationship with the government, so that the government can give us some benefits, there are many kinds of benefits, and the permission of leasehold*

Table 2 The involvement of the developer in the process of comprehensive improvement (by the authors)

	Access to projects	Returns	Developer status	Risk
Zero agent construction	Selected from a government certified pool	2–5% of total investment	Proactive	Relatively low
Open bidding	Participate in open bidding	None (may have benefits in other projects)	Passive	Relatively high

in the urban village is one of them" (interview with development company Vanke, June 2020).

The challenge is therefore, how to realise the sustainable development of comprehensive improvement for both developers and the government. As mentioned in the article of the government document above, developers are encouraged to transform urban villages in order to benefit from long-term rental apartment, which in turn provides essential affordable housing for the migrant workers that the city needs. According to interviews with developers, the initial profit preference of the developers is the "rent and urban commercial operation" of the renovated properties. The overall improvements to the environment of the whole villages also enhancing the commercial value of their investments with the resultant profit from commercial properties being their focus.

5.1.2 Negotiation

The negotiation stage is the most complex, involving many different interests. The developers in addition to having to be 'invited' to participate in the CIP, have to consider whether they want to expand into the rental apartment for profit industry. According to Article 22 and Article 18 mentioned above, the municipal government's approach to securing comprehensive improvement is to combine the needed physical improvements with the long-term rental apartment market. According to the interviews with developers, because the government is unable to undertake the construction tasks independently, the developers basically take all responsibility including negotiation with village committees regarding specific construction and other matters. Aside from favourable policies, the developers do not receive substantial support. In terms of the supervision of implementation of CIP, although the control of the government on the CIP is outlined in the 'Shenzhen Urban Village Comprehensive Management Assessment Acceptance Evaluation Standard', some problems that the government wants to solve cannot be realized. For example, Article 9 which gives priority to the rental needs of the original tenants is rarely implemented because the government does not talk directly with the tenants but only liaises directly with the developer, or through very limited contact with the village committees.

5.1.3 Implementation

The final effect of the CIP can be seen from the acceptance mechanism. The government conducted the assessment of the comprehensive improvement of urban villages according to the 'Comprehensive Management Assessment and Acceptance Evaluation Standard of Shenzhen Urban Villages (Shenzhen Municipal Bureau of Planning & Natural Resources, 2018)', which involves ten specific aspects, including sanitation, fire control, and food safety. Different projects have different scores based on their particular necessities and the difficulty of implementation. The whole projects can be seen as a success if the total score meets the requirements. During the actual construction, some necessary but difficult and costly projects will be abandoned by the developer, resulting in some hidden dangers that have not been truly solved. Also, with some projects the improvements have not been as comprehensive as the government intended and according to one interviewee: "*...sometimes, the results of the implementation are only cosmetic.*"

5.2 *Operational Issues*

The whole process of CIP has been unsustainable in many villages due to conflicting interests and the institutional imperfections of the new policy. This section relies on interviews with the parties involved in the process described above to discuss the problems encountered in the process and the potential for sustainability of the policy. The interviews questions were targeted at the different sub-sections of planning, negotiation, and implementation in an attempt to compare the real views of the parties on the same issue in parallel. The questions for the interviews are structured in the order shown in Table 3.

Table 3 Focus of the interview questions and the corresponding interviewees (by the authors)

			Government	Developer	Designer	Villager
Planning	1	Possibilities for policy promotion	●	●	●	
	2	Achievability of guidance regulations	●	●		
	3	Purpose in the CIP	●	●		●
Negotiation	4	Obstacles in communication	●	●	●	●
Implementation	5	Completion level of government-required projects		●		
	6	Current progress of the project	●	●		

5.2.1 Planning

Lack of Clarity in the Guidance Documents

Q1: Is there any possibility for the in-depth Comprehensive Improvement to be extended to other urban villages in Shenzhen on a large scale?

The government does not plan to implement the in-depth CIP approach in all urban villages, as the number of urban villages in Shenzhen are so large that some may not be suitable for an in-depth CIP approach. It will therefore be implemented according to market demand. However, there are no written rules governing this so-called market-oriented approach, and the core driver of the current deep comprehensive improvement is still the will of the government, as developers have found it difficult to find new profitable models. The developers and designers interviewed reacted negatively to the continued expansion of these projects, with the developer of Jingle Village stating that it had lost money due to an internal misjudgement of the project in the early stages. The model can be replicated, but only if the right conditions for intervention are found.

The Replacement of the Tenants Group

Q2: Will it achieve control over rents and give priority to meeting the "housing needs of the original tenants" as stated in the regulations?

The leasing objectives of the buildings renovated by developers have often changed, and it is difficult to meet the requirements of Article 9 in the Overall Plan of Urban Villages to give 'priority to the leasing needs of the original tenants.' According to surveys conducted by developers before the construction process, the main tenant groups in the urban villages are low-income groups among which families account for a large proportion. However, the long-term rental apartment is mainly targeted at young graduates. Neither the negotiation mode nor the commercial operation of the transformation can meet the requirements of giving priority to the original tenants, because the original tenants are never facilitated or desire to participate in the discussion and decision-making. As Vanke's developers say: "*Families don't rent converted flats, they think more about affordability, they don't care about en-suite bathrooms, amenities …… and they've already found somewhere else to live while you're in the process of conversion.*" This was also illustrated by interviews with tenants in the village, where one small shop owner said that he didn't like the flats and preferred a family atmosphere. He also said that: "*We don't like these flats, and if you've been inside you will know, you can't even cook in the house.*"

Apart from the fact that the type of housing available does not meet the needs of the original tenants, the increase in rents is also a key factor in deterring them from renting. In fact, from interviews with the government, we know that there is no clear mechanism to regulate rents. The government may put some pressure on the developers, but the government has no control over the rental agreement between the villagers and the developers, and it is up to the developers to negotiate, and this is where the biggest cost for the project is.

The Interest Conflicts

Q3: What do you expect to achieve from the CIP project?

The government's expectation is to solve the problems of the urban villages at a low financial cost. Significantly, there is a clear conflict of interest between the government and the developer in these projects. Unlike usual real estate development, where the price of land granted to the government is positively correlated with the price of housing, the government receives no benefit from the CIP project, except for social stability and improved sanitation, which the government expects to be self-financing. Therefore, after the government has established the project, the developer is responsible for the design, procurement, and construction. Although the cost is usually borne by the government, the government will give the developer a 2–5% of the total investment as construction fees, however, this fee does not exist if the developers are involved as 'Zero Agent Construction'. Here, the government hopes that developers will take the initiative to find a profitable point in the project to cover the base cost.

In fact, developers are not interested in construction fees, but in the secondary real estate market after the comprehensive improvement. The developer of Vanke said that: "*We want to do comprehensive improvement project because we expect that the government has helped us invest in improving the environment in surrounding area after I have done the conversion with the building', and he added: 'Very often construction is a way to make good relations with the government, and I think the purpose of this is to let the government to give us some benefits, such as the right to develop some certain projects*". However, the developers remain pessimistic about the investment cycle of the project due to the long negotiation time spent in the early stage of renovation. One developer also said that: "... *It's not until after the budget that the finance department will allocate the money. Like I said, it is basically the construction company pay their own money in advance at the early stage of the project.*" Thus, on the one hand, the government wants developers to take on more responsibility for construction and funding, while on the other hand, developers want the government to be responsible for infrastructure investment. These circumstances are an obstacle to the sustainable implementation of the policy until a more standardised implementation clause is in place.

Villagers are generally opposed to the idea of deep comprehensive improvement. According to several interviews conducted, they believe that long-term rental flats usually means the loss of control over their own buildings. Although the lease will be managed by a company that is responsible for the management of the property and the income from the rent is stable, in fact few villagers are in fact responsible for the management of the rent themselves, they usually hire someone else to manage it. They are therefore more concerned about the control of the shop units, as this is where the highest rental income is generated. The villagers therefore believe even the project can pay them well, it can still seriously infringe their interests, and this is why villagers generally ask for higher rents in return.

5.2.2 Negotiation

In the CIP process, communication is a game of chance between the interests and goals of several parties. It is clear that the government wants a low-cost solution, while the developers face a difficult negotiation situation with the villagers for their profit. Significantly, the payback period for CIP is much longer and less profitable than the traditional D+R approach.

The urban villages themselves are often the main obstacle of the implementation of CIP. Indeed, objectively speaking, construction in the urban villages is much more difficult than a normal project due to the different ownerships of the densely packed buildings, there are many things that need to be coordinated. Only very experienced developers therefore have the capability to undertake such complex development tasks. In addition, experience from previous initiatives, such as D+R, means that villagers expect large compensation sums. As a result, some villagers collectively make it difficult for the developers at the beginning of the project implementation. In Sungang village (see Figs. 2, 3 and 4), for example, villagers believed that once the CIP had started that their chances of getting compensation would be gone (revealed in an interview with a developer).

In reality, developers have no right to decide if the urban village can be demolished or not. As early as November 2018, Shenzhen Luohu municipal government promulgated the 'Implementation Measures for Urban Renewal of Luohu District, Shenzhen' which stated in Article 20 that: 'For urban villages and old housing villages where the current plot ratio exceeds 2.5, except for the need to implement major urban infrastructure and public service facilities, the renewal mode should, in principle, focus on comprehensive improvement' (Luohu District People's Government Office of Shenzhen, 2018). The negotiation process is more difficult if the developer needs to convert some of the buildings in the urban village into long-term rental apartment rather than only undertaking infrastructure improvement. In these cases, they need to negotiate not just with the village committee but with each villager individually. This usually involves the developer having to have a different team to negotiate with different building owners and since each building is different in terms of size and condition, the compensation amounts will not be the same. Moreover, some villagers have their own special demands, such as retaining one floor for their own living after the reconstruction work has been completed, which makes it more difficult for developers to conduct a unified operation.

- ***Q4: Are there any communication problems or obstacles with other parties?***

Developer—Villager

The difficulties in communication between the developer and the villagers are mainly due to the high cost of acquisition, the large number of households involved, and some of which are difficult to contact or to communicate with.

A Vanke developer in charge of Jingle Village said that the success rate of negotiations depends on the understanding of villagers and the power of the village council. The terms offered by real estate companies are generally more generous than the existing rental terms, but there is still a lot of resistance from the villagers. On the one hand, villagers are generally new to the concept of long-term rentals and are initially resistant to the idea, mistakingly believing that they will lose their chance for involvement in D+R (with extremely high compensation) if they sign a long-term lease. On the other hand, the greater the voice of the village committee, the stronger its ability to organise and publicise the facts. In this case, if the strong village committee is more supportive of enterprises entering the village, the more effective will be the propaganda and explanation for the long-term lease. Also, some village leaders take the lead in signing the contract, which has a positive demonstration effect on other villagers. If the village committee itself is more resistant to the entry of enterprises into the village at the first place or has other ideas (such as requiring multiple enterprises to compete for offers), this will have a negative impact on the villagers' wishes.

Government—Developer

According to a developer from Vanke: "*Because Jingle Village was a project new established when the comprehensive improvement was first introduced, the district government was not sure how it should be done, and it is our business that pushes the government to go through the corresponding process.*" For the government attitude on the comprehensive improvement project, he also said that:

> I feel that the government doesn't pay much attention to this kind of project. Every year the government has to have a certain number of villages to complete, and towards the end of the year he suddenly picks a unit and says: you start it. The six-month period was reduced to a few months. Then we had to work on the project construction and go through the tedious process of project approval at the same time.

In addition, interviews with designers and developers revealed that the district governments have demonstrated confusion with regard to the new policy and that there is no uniform standard to the way it is applied in different villages. Significantly, the city has no strict rules as to which department should take the responsible for the implementation of the policy. Moreover, the different departments have different specialties and different understandings of the comprehensive improvement project. For example, in Luohu district, the responsible department is the Construction Affairs

Bureau, and in Nanshan district it is the Street Office Department who implement the comprehensive improvement programme. Furthermore, some government departments are reluctant to provide funding because they think developers will be able to profit by participating in the comprehensive renovation projects, which in some cases makes it difficult to implement the programme.

5.2.3 Implementation

Hard to Achieve the Expectation

- ***Q5: What is the degree of completion of the completed CIP projects?***

It is also difficult to guarantee the results of the comprehensive improvement project so far. According to the developer, the criteria for passing the comprehensive improvement assess will be evaluated in accordance with the document 'General Command Office on the Issuance of Evaluation Criteria for the Comprehensive Management of Urban Villages in Shenzhen (2018)', some of which are very difficult to implement, he said that: "*…For example, every household is required to install a burglar net with an escape window. But few homeowners are willing to do this, fearing that the opening of the screen would be unsafe for children.*" The fact that the items in the evaluation criteria appear to be detailed, but are scored, means that many of them are not mandatory. The companies will inevitably choose to complete the project in their own favour or with an easy score. This may make it difficult for the government to achieve the desired results. Moreover, these are not projects that can obviously reflect the government's political achievements, so the government's attention is not as high as expected. It is also mentioned in previous section that some individual departments do not pay much attention to them.

Difficulties in Advancing Projects with Expansion Largely Stalled

- ***Q6: How is the project progressing now?***

In addition to the imperfect market and the lack of substantial government support, the main reason behind the lack of progress is the failure of developers in making decisions. According to interviews with designers and developers, the initial optimism about the outcomes led developers to misjudge the market, leading to high budgets being allocated for refurbishment projects. At present, there has been no team to follow up the negotiation with the villagers since the middle of 2019, and even some buildings that have completed the negotiation stage have been returned to the villagers with liquidated damages. Now only unfinished buildings are being constructed, and the operation of the renovated buildings is still continuing. Although the suspension does not mean that the project has failed, it can be said that there is no better way to make this idea sustainable at present.

However, the government officials have a different view on the current phenomenon of stagnation of integrated urban village projects, they believe that the 'invisible hand' of market will eventually balance the interest and risk. Indeed, one said that: "*It is more appropriate for the market to take the lead... These situations also have the problem of inaccurate market positioning, and some enterprises are still improving the positioning of the youth apartment in various aspects. ...if this market positioning is acceptable to the market. If not, they will certainly have to change their development strategy.*"

6 Discussion and Implications

6.1 Stakeholders' Role in the Process of Implementation

Although there are 3 + 1 different modes of participation in the implementation process (see Table 1), the government still plays the role of decision-maker in most of these modes. The government claimed in the interviews that the comprehensive approach would be carried out under the rules of the market economy, however, whether and to what extent the project can be deeply transformed needs to be approved by the government. For projects that are obviously not profitable under market rules, the government will encourage developers to take them through other interest exchanges. Therefore, the social and public welfare nature of the comprehensive improvement policy is very obvious.

While the government has played a leading and decision-making role in all aspects of the process, the active participation of developers has also been instrumental. Although not as lucrative as in the era of D+R, developers are still actively trying to find new ways to break through in the early stages of CIP. In the case of successful projects such as the redevelopment of Jingle Village, the response from the community has been positive. Although the profitability of the project has been modest, its performance in the community has largely achieved the original purpose of the project, particularly in terms of environmental improvements.

The role of the villagers is the same as in the previous D+R approach, with them still acting as landlords, however, without the intense social conflict generated by the previous strategy (Hin & Xin, 2011). Nevertheless, the villagers had some concerns about their control of property rights and expectations of profitability, this only determined the level of depth of the comprehensive improvement and did not generate further social conflict.

6.2 Issues in the Process of Implementation

During the planning stage of the implementation process, problems are focused on the lack of clearer implementation guidelines, the lack of guidance on the implementation of specific objectives, and the lack of mediation measures to address conflicting interests. The lack of clarity about the government's expectations and implementation details makes it difficult for developers and villagers to make accurate judgements about the performance of the project. The developer's representatives clearly indicated that they would not invest more in the in-depth comprehensive improvement project for the time being, and the 'uncertainty of the scope of government investment' and the 'higher than expected operational and management costs' made them doubtful and hesitant about the project's expectations. Some of the social targets and expectations for the project are difficult to meet, such as the return of former tenants and the control of rental costs. The conflict of interest in the project has resulted in the failure to implement many social ambitions, which requires further collaboration between the government and the developer to find a balance between investment and profitability.

During the negotiation phase, the main problem is still related to the conflict of interest, but more notable is the lack of understanding and misunderstanding of the policy. As some villagers had previously received large amounts of compensation for D+R, they saw the new policy as an infringement of their interests. In fact, the government has strict criteria for the possibility of D+R, and the old strategy has been criticised widely by the society. Nevertheless, in some villages, regeneration has been implemented relatively smoothly, on the one hand because of protection measures such as generous lease agreements, which guarantee the rights and interests of villagers, and on the other hand because the government and the village council have strengthened their cooperation in explaining policies to villagers. Although the issue is essentially one of interest, the government's attention and involvement and the high level of negotiation can appropriately reduce villagers' confrontation.

In the implementation process, there are two main problems, the degree of completion of the project and the motivation to implement it. Leaving aside the situation of in-depth comprehensive improvement, there are problems with the evaluation criteria and the lack of attention from the various departments in the basic CIP. The balance between the priority projects and the difficulty of implementation needs to be further studied and improved, in order to achieve better project expectations. The high cost of implementation is also the main reason for the stagnation of the project, as the developer was overly optimistic about the project's profitability, which led to a misjudgement of the project's criteria and investment, thus overstretching the project's ability to continue to expand in the short term. As such, developers need to optimise their business model before they can find better profitability.

7 Conclusions

It is evident from the interviews with the various stakeholders, CIP is not a single strategy but a series of measures. Indeed, the basic measure is to ensure the optimisation of the physical conditions within the urban village, while the depth of the comprehensive improvement will have a different urban performance in different villages. In addition to the UMV model, other cases involve several stakeholders. The government has played the role of guide and decision maker, taking into account the development of the city and social harmony. There is, however, still a need to strengthen the interpretation of the regulations, moreover, to strengthen the coordination and financial support of the government in the key aspects of implementation. Indeed, to date, the developers have put a lot of effort and capital into the initial exploration of the projects thereby making a considerable contribution to the exploration of the project on a trial basis. Nevertheless, the investment in the projects needs to be better analysed in terms of feasibility to maintain the ability to sustain the project. Also, the villagers concerns about their interests have hindered implementation during the negotiation phase, but as the most affected stakeholders, they need better communication approach and policy clarification, and this needs to be further strengthened by the government and the developer. Of these, the role of the developer is the most crucial, as it is involved in almost every aspect of implementation, and commercialisation is also key to the sustainability of the policy. This research therefore demonstrates that better coordination between the various stakeholders will further contribute to the sustainable development of urban regeneration in informal settlements such as urban villages.

References

Bromley, R. D. F., Tallon, A. R., & Thomas, C. J. (2016). City centre regeneration through residential development: Contributing to sustainability. *Urban Studies, 42*(13), 2407–2429. https://doi.org/10.1080/00420980500379537

Chiu-Shee, C., & Zheng, S. (2019). A burden or a tool? Rationalizing public housing provision in Chinese cities. *Housing Studies, 36*(4), 500–543. https://doi.org/10.1080/02673037.2019.1667490

Cybriwsky, R. (2013). The new urban frontier: Gentrification and the Revanchist City, Neil Smith. *Urban Geography, 20*(1), 93–94. https://doi.org/10.2747/0272-3638.20.1.93

Davis, M. (2006). Planet of slums. *New Perspectives Quarterly, 23*(2), 6–11. https://doi.org/10.1111/J.1540-5842.2006.00797.X

Doshi, S. (2013). The politics of the evicted: Redevelopment, subjectivity, and difference in Mumbai's slum frontier. *Antipode, 45*(4), 844–865. https://doi.org/10.1111/J.1467-8330.2012.01023.X

Dovey, K., van Oostrum, M., Chatterjee, I., & Shafique, T. (2020). Towards a morphogenesis of informal settlements. *Habitat International, 104*, 102240. https://doi.org/10.1016/J.HABITATINT.2020.102240

Dowall, D. E. (1991). Comparing Karachi's informal and formal housing delivery systems. *Cities, 8*(3), 217–227. https://doi.org/10.1016/0264-2751(91)90061-U

Dupont, V., Jordhus-Lier, D., Sutherland, C., & Braathen, E. (2015). *The politics of slums in the global south: Urban informality in Brazil, India*. Routledge.

Fekade, W. (2000). Deficits of formal urban land management and informal responses under rapid urban growth, an international perspective. *Habitat International, 24*(2), 127–150. https://doi.org/10.1016/S0197-3975(99)00034-X

Gao, Y., Shahab, S., & Ahmadpoor, N. (2020). Morphology of urban villages in China: A case study of Dayuan Village in Guangzhou. *Urban Science, 4*(2). https://doi.org/10.3390/urbansci4020023

Hin, L. L., & Xin, L. (2011). Redevelopment of urban villages in Shenzhen, China—An analysis of power relations and urban coalitions. *Habitat International, 35*(3), 426–434. https://doi.org/10.1016/J.HABITATINT.2010.12.001

Huchzermeyer, M. (2011). *Cities with 'slums': From informal settlement eradication to a right to the city in Africa*. Juta and Company Ltd.

Jiang, Y., Mohabir, N., Ma, R., Wu, L., & Chen, M. (2020). Whose village? Stakeholder interests in the urban renewal of Hubei old village in Shenzhen. *Land Use Policy, 91*. https://doi.org/10.1016/j.landusepol.2019.104411

Kriese, U., & Scholz, R. W. (2011). The positioning of sustainability within residential property marketing. *Urban Studies, 48*(7), 1503–1527. https://doi.org/10.1177/0042098010375321

Li, B., Tong, D., Wu, Y., & Li, G. (2021). Government-backed 'laundering of the grey' in upgrading urban village properties: Ningmeng Apartment Project in Shuiwei Village, Shenzhen, China. *Progress in Planning, 146*, 100436. https://doi.org/10.1016/J.PROGRESS.2019.100436

Liu, Y., Geertman, S., van Oort, F., & Lin, Y. (2018). Making the 'invisible' visible: Redevelopment-induced displacement of migrants in Shenzhen, China. *International Journal of Urban and Regional Research, 42*(3), 483–499. https://doi.org/10.1111/1468-2427.12646

Liu, Y., Wu, F., Liu, Y., & Li, Z. (2016). Changing neighbourhood cohesion under the impact of urban redevelopment: A case study of Guangzhou, China. *Urban Geography, 38*(2), 266–290. https://doi.org/10.1080/02723638.2016.1152842

Luohu District People's Government Office of Shenzhen. (2018). *Notice of Luohu District People's Government Office of Shenzhen Municipality on printing and distributing the Action Plan for comprehensive Management of "Villages in Cities" in Luohu District (2018–2020).*

Pan, W., & Du, J. (2021). Towards sustainable urban transition: A critical review of strategies and policies of urban village renewal in Shenzhen, China. *Land Use Policy, 111*, 105744. https://doi.org/10.1016/J.LANDUSEPOL.2021.105744

Ren, X. (2017). Governing the informal: Housing policies over informal settlements in China, India, and Brazil. *Housing Policy Debate, 28*(1), 79–93. https://doi.org/10.1080/10511482.2016.1247105

Shenzhen Municipal Bureau of Planning and Natural Resources. (2018). *What is the urban renewal of comprehensive improvement?* http://www.sz.gov.cn/ghj/ywgz_3_3/qtl/201811/t20181115_14516897.htm

Shenzhen Municipal Bureau of Planning and Natural Resources. (2019). *Overall planning for comprehensive improvement of urban villages (old villages) in Shenzhen.*

Tan, Y., He, J., Han, H., & Zhang, W. (2019). Evaluating residents' satisfaction with market-oriented urban village transformation: A case study of Yangji Village in Guangzhou, China. *Cities, 95*(July). https://doi.org/10.1016/j.cities.2019.102394

Uehara, Y. (2008). Unknown urbanity: Towards the village in the city. *Architectural Design, 78*(5), 52–55.

Wang, Y. Y. P., Wang, Y. Y. P., & Wu, J. (2009). Urbanization and informal development in China: Urban villages in Shenzhen. *International Journal of Urban and Regional Research, 33*(4), 957–973. https://doi.org/10.1111/J.1468-2427.2009.00891.X

Wang, Y. P., Wang, Y., & Wu, J. (2010). Housing migrant workers in rapidly urbanizing regions: A study of the Chinese model in Shenzhen. *Housing Studies, 25*(1), 83–100. https://doi.org/10.1080/02673030903362019

Wang, H., Zhao, Y., Gao, X., & Gao, B. (2021). Collaborative decision-making for urban regeneration: A literature review and bibliometric analysis. *Land Use Policy, 107*, 105479.

Wu, F. (2016). Housing in Chinese urban villages: The dwellers, conditions and tenancy informality. *Housing Studies, 31*(7), 852–870. https://doi.org/10.1080/02673037.2016.1150429

Wu, F., Zhang, F., & Webster, C. (2012). Informality and the development and demolition of urban villages in the Chinese peri-urban area. *Urban Studies, 50*(10), 1919–1934. https://doi.org/10.1177/0042098012466600

Xinhua News Agency. (2006). *Footprint of the Republic—2006: Building a harmonious society*. http://www.gov.cn/test/2009-10/20/content_1444026.htm

Yangteng, Y. (2004). Rebuilding villages in the city, cutting "tumors" in Shenzhen. *The Economic Journal*.

Zhang, L., Zhao, S. X. B., & Tian, J. P. (2003). Self-help in housing and chengzhongcun in China's urbanization. *International Journal of Urban and Regional Research, 27*(4), 912–937. https://doi.org/10.1111/J.0309-1317.2003.00491.X

Zheng, H. W., Shen, G. Q., & Wang, H. (2014). A review of recent studies on sustainable urban renewal. *Habitat International, 41*, 272–279. https://doi.org/10.1016/j.habitatint.2013.08.006

Zhou, B. (2004). *Sigh of shenzhen police* (pp. 6–12).

Part IV
Participation in Ecological Regeneration

Chapter 8
An Ecology Model for Participatory Strategies: Community-Led Green Networks and Its Social and Spatial Agents

Diana M. Benjumea Mejia

Abstract Over the past three decades, the city of Medellin has sustained dramatic transformations, transitioning from being one of the most violent cities in the 1990s to emerging as one of Colombia's most resilient, innovative, and greenest urban centers. A key component driving these changes has been the revitalization of lower-income neighbourhoods through participatory strategies aimed at reducing violence, enhancing social resilience, and promoting inclusivity via urban green projects. A remarkable initiative was the Green Belt project included in the Plan Maestro Area Centroriental, designed to connect the most relevant nodes through walking and cycling paths, recreational facilities, and diverse urban green infrastructure such as urban farming, community gardens, parks. Government-led participatory strategies have not only beautified the neighbourhoods but have also garnered both local and international attention. Nevertheless, the shortcoming of these strategies in Medellin is the fragile relationship between stakeholders and spatial ownership, particularly concerning green spaces. Conflicts occur when green areas previously managed by communities, are transferred to the local governments, fracturing both activism and spatial equality. Amidst these deliberations, the significance of an overlooked green infrastructure that has survived for decades is the small-scale community-led green interventions often located at the fringes of residual green areas (i.e., nearby roads or areas next to water streams). The random appropriation of residual green spaces depicts vital signs of pro-environmental behaviours, reflecting commitments to nature preservation and environmental agency. This green network, acknowledging the involvement of both human and non-human living agents (e.g., plants, trees, and wildlife), is often overlooked in traditional urban design and planning approaches. This chapter discusses the potential of these often disregarded small-scale community-led green networks as complementary avenues

D. M. Benjumea Mejia (✉)
Centre for Core Learning, College of Interdisciplinary and Experiential Learning, Singapore University of Social Sciences, Singapore, Singapore
e-mail: arqdmbm@hotmail.com; dianabenjumea@suss.edu.sg

Representation, Communication, and Architecture Research Group, National University of Colombia, Manizales, Colombia

E. Mangi et al. (eds.), *Innovative Public Participation Practices for Sustainable Urban Regeneration*, Urban Sustainability, https://doi.org/10.1007/978-981-99-9595-0_8

to government-led participatory strategies in urban green revitalisation projects. Through an exploration of the diverse living agents involved in this processes, the levels of interventions, and modes of autonomous participation, the chapter unravels and analyses these aspects to formulate an ecology of participation model. This model introduces a paradigm emphasising equitable interactions in urban green projects, grounded in a living ecology framework that recognises the social and political agency of all spatial actors.

1 Introduction

The urban population in Latino America has experienced exponential growth, reaching an approximate 37% increase over the last three decades (World Bank, 2022). The unpreparedness of cities and insufficient infrastructure have intensified challenges in the main urban areas. Climate change, pollution, urban sprawl, crime, and poverty are some of the most visible problems arising in the past years. Proposed solutions have predominantly focused on three main aspects: (1) climate resilience through urban green infrastructure; (2) crime reduction programs; (3) inclusive and pedagogical participatory processes of lower income communities. A noteworthy example of this three-fold strategy is the city of Medellin (Colombia). Over the past thirty years, Medellin has transformed from a city beset with territories of fear caused by crime and drug trafficking, to one praised as the most resilient city to crime (Resilience Office, 2016), most innovative (Urban Land Institute, 2013), and a city that celebrates life (Lee Kuan Yew, 2016). Part of these recognitions are owed to the city rebranding programs aimed at investing in high profile infrastructure that include lower income settlements (Franz, 2017; Sotomayor, 2017). Through Social Urbanism (SU)—a government-developed participatory scheme that promotes inclusive urbanism in lower income neighbourhoods (Benjumea, 2020), Medellin has enabled social empowerment across its Valley.

Over the past two decades, SU has been the main local government's flagship to reduce crime and improve local infrastructure. One of the most resent and ambitious projects is The Metropolitan Greenbelt, conceived to control urban sprawl of lower income settlements. This project which transforms the city borders into urban green spaces (Echeverry Tamayo, 2019) spans the eastern and western mountains, encompassing a series of interventions that provide urban green services and improve the natural, social, and financial environments of the city. The Greenbelt targets climate change regulations, leisure spaces, urban agriculture, and environmental restoration of native ecosystems (Anguelovski et al., 2019). Included in the Greenbelt is the Circumvent Garden (Jardin Circunvalar), a pilot project built in 2016 at the North-eastern part of the city in the lower income area known as Comuna 8. Collaboratively designed with residents, the Circumvent Garden employed educative and participatory strategies that engaged residents at different stages of the project (e.g., diagnosis, design, and construction). The project comprised a total of 206,305 m^2 transformed into public space, three eco-parks and environmental classrooms, 30 ha of environmental restoration, and 31,000 m^2 of community agroecological gardens (OPPCM, 2016).

The integration of participatory principles in the Circumvent Garden has positioned the project as an international exemplar (Lee Kuan Yew, 2016). Nevertheless, the conventional approach to infrastructure design and provision of urban services was largely dominated by centralised grey infrastructure exposing gaps in the planning, design, and management of the project. Notably, the oversight of urban ecosystem services (ES) including urban vegetation, provision of non-invasive native vegetation, holistic and adequate design of green infrastructure to provide cooling, shading, water absorption, and carbon sequestration, has been evident. Furthermore, the cultural and social dimensions also revealed shortcomings in the participatory process. In contrast to the SU tradition, where community-based experts are actively involved in the process, project experts worked in isolation and transferred control of the green spaces to private organizations (Anguelovski et al., 2019). The marginalisation of local environmental stewards and their community-led nature interventions resulted in disengagement of the communities in environmental processes (Benjumea and Olszewska-Guizzo, 2021). This highlights equity and environmental-justice issues, particularly concerning urban green gentrifications[1] in which the biophysical and the previously common resources are politically negotiated, compromised, and privatized at the expense of local communities (Anguelovski et al., 2019; Sotomayor, 2021).

Considerable discourse surrounds participatory models and their attendant challenges. The mediation among stakeholders during planning, management, and physical interventions unleashes socio-physical, political, and economic changes at the local and city levels (Varley, 2013). Challenges arise when environmental equity and justice are sought in interventions emphasising agroecological parks (e.g., community farming) and environmental restoration in lower income settlements, exemplified by the Circumvent Garden. The scope of such projects should extend beyond the integration of ecosystem services (ES) principles encompassing social, ecological, and technological dimensions focused solely on the provision of urban green infrastructure. Instead, it should acknowledge the roles of key agents and relative contributions that shift in time and space thereby influencing dynamic changes across time (McPhearson et al., 2022). Furthermore, the conventional ES tend to overlook the environmental agency exhibited by non-professionals or community-experts establishing networks of interaction in residual urban green spaces (Benjumea et al., 2022). These networks introduce a subtle participatory dimension that considers all non-human living agents—such as plants, trees, and wildlife—as fundamental to sustain healthy urban ecosystems. This approach allows the integration of vegetal

[1] Green gentrification is defined as the displacement or exclusion of vulnerable populations while exposing environmental ethics through green initiatives that push low income, working class, and minority residents away (Dooling, 2009; Gould & Lewis, 2017). In the case of Medellin Circumvent Garden Anguelovski and colleagues (2019) observed that the environmental actions of low-income residents were replaced by a transformed and newly valued nature, a clear form of urban green gentrification.

politics in urban-green spatial design processes (Elton, 2021). The significance of these networks lies in their contribution as cultural ES representing non-material benefits derived from people's interactions with nature, recognized as highly valued (Camps-Calvet et al., 2016).

Consideration of non-human networks in the planning and design of urban green spaces might constitute a challenge in cities located in developing countries. This is due to the fragmented urbanism that emerges with new urban green projects often lacking regulatory planning laws and strategies of urban integration. This planning approach also reveals a deficiency in system thinking, overlooking at the interrelationships and dependencies between critical resources e.g., natural, socioeconomic, and cultural (Frank et al., 2017). The relational aspects between systems prompts proposals that not only address complex problems through different frameworks but also bridge the dichotomy between the built and non-human environment (Descola, 1996). Some renowned frameworks have integrated principles of social practice theory (Lach et al., 2022). Departing from the premise of socially understanding the normative activities and culturally relevant practices, social practice theory recognizes three predominant dimensions: skills, meanings, and materials (Lach et al., 2022). For instance, any interaction with an object would require a know-how to manipulate it, followed by a meaning or symbol attached to it that might evoke certain feeling or sense of comfort. While this framework proposed by Lach and colleagues places community wellbeing and community participation at its core, it does not entirely overcome the dominat human-centric view prevalent in ES urban practices wherein nature is often regarded as a commodity for human use (Ives et al., 2018). In this context, nature is positioned as the object of interaction and/or mediator to evoke human meanings in the local space.

Overlooking at plants agency within any participatory framework might be as critical as overlooking the agency of communities. Approaches in this arena, stand to benefit from integrating ecology principles into practices that: "*would ground human intention and action within the context of an ongoing and mutually constitutive engagement between people and their environments*" (Inglod, 1996: 26). Dekay proposes an approximation of spatial design from a system thinking perspective that consider ecological principles emphasising actions based on ecological values—a practice that "*understands the interrelatedness of life*" (Dekay, 1996, p. 4). However, the systems thinking perspective should extend beyond the objective notion of structural functionality and its human-established values. The ecology living framework proposed in this chapter considers the integration of asymmetrical relationships between humans and non-humans, determining their participation and representation in space, including its temporal variations, compatibilities, and incompatibilities between living agents (Descola, 1996). Drawing inspiration from established ecology models in ecological psychology (Gibson, 1979), geography and health and wellbeing (Irvine et al., 2019; Ives et al., 2018), community stewardship and participation

(Benjumea et al., 2022), plants politics (DelSesto, 2020; Marder, 2012) and anthropology (Kohn, 2013), the framework evolves into four interrelated elements: Entanglements and Equity Cooperation; Spatial–temporal representation (**ECOS**). The subsequent sections present findings from a study conducted since 2016 in various neighbourhoods in Medellin. A comparison between community-led gardens and the Circumvent Garden project unfolds, elucidating different spatial actors and socio-political entanglements in the participatory arena. The discussion section introduces the ECOS framework offering suggestions for future implementation and debates.

2 Methods

A within comparative case study was conducted since 2016 in the comuna 4, Aranjuez; comuna 13, San Javier; and Comuna 8, Villa Hermosa (see Fig. 1). Community-led green spaces were systematically explored and analysed following a comparative principle within the context of their respective Neighbourhoods. This approach facilitated the identification of iterative and contingent relevant factors in the social and physical space, unveiling the processes that gave rise to community-led nature initiatives. To gain insights into the spatial and social characteristics of the green spaces and its impact in residents' perception of interrelatedness with the natural environment, interviews (N = 30) and surveys were conducted with residents (N = 200) and local leaders (N = 10) actively contributing to local green projects. Spatial assessments were performed in the community-led spaces and the circumvent garden (N = 3 respectively). The assessment criteria were based on a system of ecological vegetation units evaluated against specific structural characteristics (Daniels et al., 2018), such as water elements, lawns, flower beds, hedges, and margins assessed against the ecological properties of the space, namely, element quality, ecological integrity, habitat animals, habitat plants, and pollination (IBID). These were expert-evaluated during field observations.

3 Findings

Different space categories were identified according to residents, engagement processes, typology of green space, and type of agency within the space. Three social components crucial in defining the interrelatedness with the natural environment and fostering ecological processes in these green spaces were identified: prosocial behaviours, personal interconnectedness with nature, and defence of the territory.

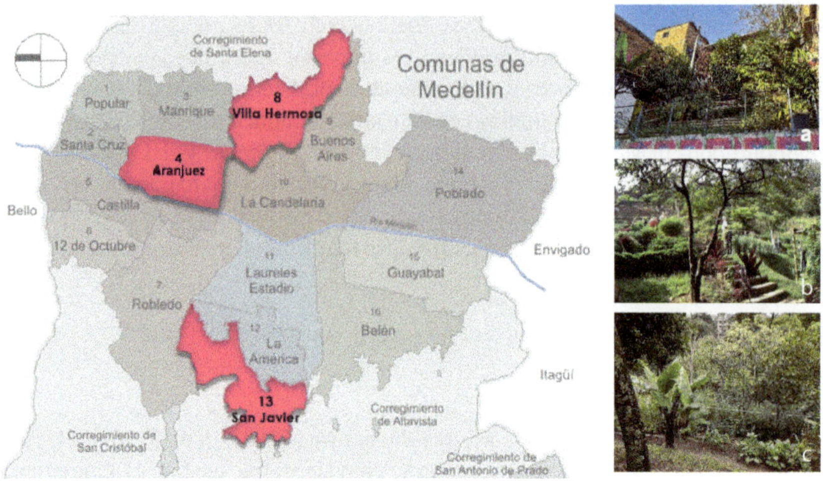

Fig. 1 **a** Comuna 4 Aranjuez; **b** Comuna 8 Villa Hermosa, and **c** Comuna 13 San Javier (*Photos* Author's private archives)

3.1 Space Categories: Communal

Communal-led gardens emerge as spaces created by residents, frequently repurposing residual or depleted urban areas. These endeavors are facilitated with the support of local leaders who stream funding sources through the local government or other organizations. The maintenance of the spaces hinges on internal community networks. One key aspect of these initiatives is the level of activism and pro-environmental behaviours that are enacted in those participating in the green spaces. In terms of its ecological contribution, these spaces are often non-curated with a wide variety of plants and biodiversity (Fig. 2a). An additional facet is the educational dimesion introduced by some of these spaces, incorporating activities such as gardening, urban farming, permaculture among others.

3.2 Private

Private initiatives manifest within the confines of resident's private land. In some instances, these initiatives adopt a mixed-economy model in which landowners derive modest financial gains from their produce, or engage in exchanges with other landowners. Although these initiatives occur in private land, it was observed that some of these spaces were originally public areas later privatised by residents. While official ownership over the land may not be established, the spaces exhibit restricted access (refer to Fig. 2b). The foremost contribution of these initiatives lies in safeguarding green areas from environmental degradation.

Fig. 2 **a** Private initiatives are developed within resident's private land; **b** educational programmes offered in community-led green spaces; **c** initiatives in public spaces (*Photos* author's private archives)

3.3 Public

These initiatives occur in residual public spaces primarily utilised for planting flowers. Following the onset of the Covid-19 pandemic, an increase in such initiatives using residual spaces for urban farming purposes was observed. According to the interviews conducted, these initiatives garnered support from government agencies in the form of seeds, gardening tools, and utensils (Fig. 2c).

3.4 Social Components

The analysis revealed three underlying social components that define the interrelatedness of residents with the natural environment in the green spaces: prosocial behaviours, personal interconnectedness with nature, and defence of the territory. Unlike the spaces within the Circumvent Garden, community-led spaces were considered as venues for meaningful and spontaneous interaction with nature and neighbours. These spaces are created and maintained by residents, providing flexibility in the range of activities that can be performed. This differs from government-built green spaces that restrict the function and ecosystems allowed within. The flexibility inherent in community-led spaces facilitates a direct experience with nature, creating opportunities of enacting prosocial behaviours geared toward the protection of the natural environment. This endavour is further reinforced by educational projects directly overseen and managed by the communities.

Interconnectedness with nature is emotionally and spiritually defined by feelings of attachment or empathy towards nature (Ives et al., 2018) and the feeling of oneness with nature (Irvine et al., 2019). Stewards of community gardens in Medellin and other urban settings (Bell-Williams et al., 2021), emphasise that gardens enable a meaningful relationship with nature, where plants and animals are viewed as an extension of themselves and families (IBID). This feeling of familiarity was found to promote defence of the territory instigating protection against illegal and criminal activities or abusive behaviours towards the natural environment such as littering and deforestation.

3.5 Spatial Assessments Ecological Properties

The spatial evaluation, of structural elements, from the ecological perspective between community and government-led green spaces revealed significant differences in the design and ecological contributions for nearly every criterion (structural and ecological) displaying differences ranging from 2.0–2.5 points, with the exception of water elements. The most pronounced differences are observed in ecological quality, habitat of plants, and pollination. Structural characteristics in community-led appeared to support a variety of ecological properties, as indicated in Table 1 and Fig. 3. Notably, the presence of natural flower beds is associated with an enhancement in pollination performance and the habitat of insect-pollinated plants and pollinators (Daniels et al., 2018) (Table 2).

4 Discussion

The findings of this study indicate that interactions in urban green spaces led by community members play a crucial role in fostering prosocial behaviours and interrelatedness with the natural environment. This phenomenon is likely attributed to the heightened ecological contribution and direct experience of natural environments (Benjumea et al., 2022). Although government-led green spaces, exemplified by the Circumvent Garden, received a lower ecology evaluation, these places were frequently lauded as aesthetically appealing and perceived as enhancements to the neighbourhood image. A relevant social component influencing the dynamics of these spaces lies in the type of human-nature connection occurring in the spaces. Community-led gardens enact a fundamental human level of interdependence with nature, a characteristic potentially derived from the flexibility inherent in creating and

Table 1 Assigned indicator average values ecological properties government-led versus community-led green spaces

Structural element	Typology	Ecology (expert knowledge max 5.0)				
		Element quality	Ecological quality	Habitats animals	Habitat plants	Pollination
Water element	Gov.	2	2.5	1.5	1.5	1
	Com.	2	2.5	1.5	1.5	1
Lawn	Gov.	2	2	2	2	2
	Com.	4	4	2.5	4.5	4
Flower bed	Gov.	2	2	2	2.5	2
	Com.	3.5	4	2	4	4
Hedges	Gov.	2	2	2	2	3
	Com.	4.5	4	3	4.5	4.5
Margin	Gov.	2	2	1.5	2	2
	Com.	4	4	3	4	4

Table 2 Unweighted means, minimum and maximum of the indicator values for ecological assessment

Structural element	Typology	Ecology means (min/max)	Absolute difference
Water element	Gov.	1.62 (1.0/2.5)	0
	Com.	1.62 (1.0/2.5)	
Lawn	Gov.	2.37 (2.0/3.0)	1.4
	Com.	3.72 (2.5/4.5)	
Flower bed	Gov.	2.09 (2.0/2.5)	1.4
	Com.	3.39 (2.0/4.0)	
Hedges	Gov.	2.16 (2.0/3.0)	1.9
	Com.	4.05 (3.0/4.5)	
Margin	Gov.	1.88 (1.5/2.0)	1.9
	Com.	3.77 (3.0/4.0)	

participating in these spaces, allowing residents to establish a profound connection with nature and assume roles as stewards of these environments.

Increasing the availability of green spaces in the city of Medellin is crucial, hence, the Circumvent Garden marks a noteworthy stride in introducing ES in lower-income neighbourhoods. Nevertheless, it is important to underscore that merely increasing the number of green spaces is not enough to instigate societal changes over the preservation and respect towards the natural environment. Therefore, focusing soley on implementing superficial interventions is unlikely to yield substatial systemic changes. Contrastingly, interventions that introduce a deeper approach of ES, encompassing the cultivation of meaningful connections to nature

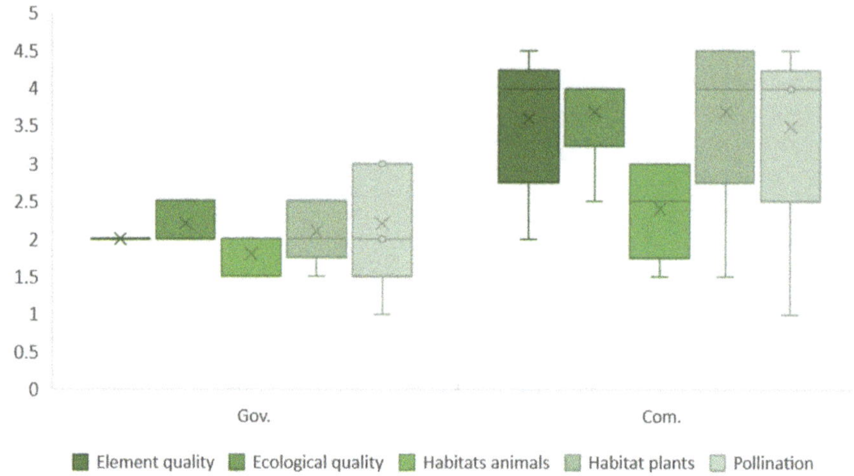

Fig. 3 Values comparison for ecological assessment between government-led and community-led green spaces

through collective and individual participation, as exemplified by the community-led green spaces, could bring positive transformations in urban contexts with depleted or limited natural environments (Ives et al., 2018). Namely, prosocial behaviours, personal interconnectedness with nature, defence of the territory and sustainability.

A contributing aspect of this study is to propose an ecology model for spatial design that provides alternatives to development, aiming to address or ameliorate fractured human-nature relationships. Creating the conditions for humans thriving in urban environments should also consider all living organisms that take part in the process. Aligned with Gibson's ecological model of human perception, at its core, ecology, involves all organisms moving in their environment, creating relationships and interdependencies or mutualism, as organisms and their environments are inherently interconnected and co-dependent (Gibson, 1979). An ecology model holds the potential to overcome the increased exploitation and indiscriminate commodification of the natural environment and consequential social inequalities evident in urban green spaces, such as degraded and devalued spaces, tensions between space users, gentrification, and degradation of biodiversity by tourism (Hunter et al., 2022).

The proposal is to shift concepts of transactional relationships between the natural and human environments in urban contexts, towards a vision of **symbiotic interrelatedness**. This conceptual framework acknowledges the human self as intricately interwoven with the living ecology of plants and animals, aligning with the perspective articulated by Koh (2013). Arne Naess expounded upon this symbiotic interrelatedness, framing it as an integral component of an ecological self. From his view, *"self, subject, object and environment are part of an indivisible unit"* (DelSesto,

2020) that operates in accordance to emerging patterns of movement that determine how an environment is occupied and lived (Gibson, 1979). These movements are multileveled and contingent upon the temporal and frequency of contact with a specific energy of action (e.g., interaction or connection with the natural environment) that derives an inner sensitivity with the action performed. A subsequent level deals with the social and community inter-relatedness. Tangible negotiation tools are formed to define rules of action that benefit both social groups and non-human organisms. These rules, also proposed by Gibson as affordances, are in direct opposition to the conventional sociological view of utilizing natural and built elements to benefit human experience (Lach et al., 2022). Instead they organically evolve directly correlated with natural patterns that define the transitions and adaptation between situations that dictate new actions. This iterative process rules the symbiotic inter-relatedness between the systems involved allowing for the emergence of new patterns and rules of adaptation. Such adaptability ensures appropriateness in time and space, accommodating different situations tied to socio-economic or political patterns as they arise.

Figure 4 introduces the ECOS conceptual framework comprised of four inter-related elements: Entanglements and Equity (i.e., inclusivity through symbolic human-nature interrelatedness); Cooperation (i.e., stakeholders, participation, plants agency); Organization (i.e., stewardship, activism, and systems functionalities); Spatial–temporal representation (i.e., structural elements) (**ECOS**).

4.1 Entanglements and Equity

At this level, autonomous participation within communities undergoes a transformative process, evolving into a shared historical continuum over time, outliving narratives, practices, and repertoires shaped by individuals (Barthel et al., 2010). The learning process and meaning-making is not static, and is characterised by reflexibility and adaptation to varying circumstances (Lawrence, 2009). The emergent practices form a social memory, serving as a source of community resilience (Folke et al., 2003). This participatory mode is often found in community-led green spaces, which emerge organically rather than through a top-down planning process, initiated through *ad-hoc* gardening endeavours (Camps-Calvet et al., 2016). In these initiatives, groups engage in ecosystem management activities such as gardening, farming, creating **social-ecological memories** and values that are "retained, modified, revived, and transmitted through time" (Barthel et al., 2010: 256; Langemeyer et al., 2018). The acquired knowledge is passed down by elders or community members possessing vast knowledge in ecosystem management, often thorugh a history of engagement in farming activities in the countryside.[2] The action of individually and socially connecting to the land and nature through practice and knowledge exchange, suggests

[2] The last four decades, thousands of countryside farmers scattered around the country in a painful diaspora caused by the civil war. Thousands of countryside farmers migrated to Medellin seeking

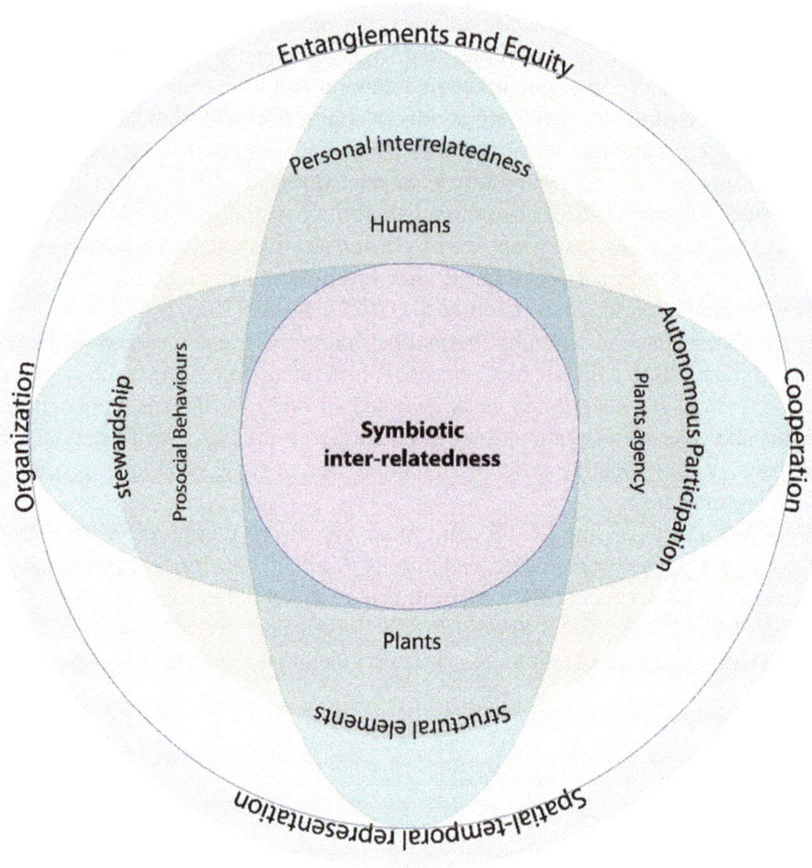

Fig. 4 ECOS conceptual framework

a profound connection between both human and non-human agents. This connection extends beyond mere collaboration and cooperation, invoking a sense of kindship and an intimate relation between humans and plants (Elton, 2021).

4.2 *Cooperation*

Engaging with non-human life entails to understand its processes, needs, and spatial requiriments. According to Marder (2012), the life of vegetal beings is "de-centred"

---

refuge and opportunities. Many of them continued with ecosystems management traditions in urban settlements.

and disseminated throughout the "body of plant communities" (Elton, 2021). Additionally, these beings contribute to place-making in the form of interconnected networks (Marder, 2012). In fact, it has been argued that the relationship between humans and plants affords the plants their spatial agency, as both are entangled in politically significant ways (Lumber et al., 2017; Mayer & Frantz's, 2004). The establishment of networks among different stakeholders is fundamental to forge participatory dynamics that are respectful with nature and consider residents as interconnected agents within the green spaces. The benefits of cooperating with nature in participatory processes yields several benefits to humans, particularly on a psychological level. These include an increased subjective feeling of wellbeing, experience oneness, wonder, unity, inspiration, awe, and a sense of harmony with the broader creation (Irvine et al., 2019). Furthermore, such collaborations foster intrinsic motivation and sense of community (Benjumea et al., 2022).

4.3 Organization

Feeling interdependent and interrelated with the rest of the natural world serves as a catalyst for ecological activism and can result in biodiversity conservation, also demonstrating an increased level of organisation. Studies have found that feeling connected to nature can instigate societal transformations related to environmental stewardship and sustainability (Ives et al., 2018; Langemeyer et al., 2018), bolster social cohesion, foster a sense of place and community (Benjumea et al., 2022), and facilitate urban residents in reconnecting with nature (Krasny & Tidball, 2009). The community-driven organization to breathe life into green spaces aligns with a fundamental principle of environmental contribution and reciprocal exchange with nature.

4.4 Spatial–Temporal Representation and Design

This level contemplates the tools that enable interactions and representations in space. At its core is the notion of **symbiotic inter-relatedness** with nature and its living components, which are not seen as independent objects but fundamental and constituting active agents in the process of representation and reproduction of space. The structural elements necessary for this symbiosis in urban space must adhere to design principles that respect the natural environment. This implies ensuring that ecological processes can thrive in the physical space, transcending shallow interventions. Considering the ecology of plants and its agency in the design of urban green spaces is the first component to ensure that a truly participative process. From this perspective, the notion of participation disengages from the conventional practise-based tools often employed in urban and planning frameworks, urging designers and urban planners to adopt alternative design paths that are attuned to both human and vegetal life. Essential to this approach is active listening to the voices of community and residents, engaging in a meaningful dialogue to understand their needs and aspirations. Importantly, the same principle should be extended to vegetal agents present in the physical space. Regular surveying and ecological evaluation of natural systems are integral to foster the process of **symbiotic inter-relatedness**.

5 Conclusions

This chapter introduces an ecology of participation model centered on small-scale community-led green spaces within a paradigm of equitable interactions in urban green projects, grownded in a living ecology framework. The framework acknowledges the different living agents that intervene in the process, the various levels of interventions, and forms of autonomous participation, categorising them into four interrelated elements: Entanglements and Equity (i.e., inclusivity through symbolic human-nature interrelatedness); Cooperation (i.e., stakeholders, participation, plants agency); Organization (encompassing stewardship, activism, and systems, functionalities); Spatial–temporal representation and design (focusing on structural elements). The consideration of the social and political agency of all spatial actors extends not only to humans but also to plants, recognizing their interconnectedness. The proposition emphasizes the significance of meaningful interconnectedness with living organisms as an underlying aspect to enact prosocial behaviours, environmental awareness, sustainability, and environmental stewardship. At its core, this proposal seeks to advocate for alternative approaches in the design of urban spaces in lower-income areas, acknowledging the value of small-scale green interventions with heightened ecological impact as integral components of secondary green networks.

References

Anguelovski, I., Irazabal, C., & Connolly, J. J. T. (2019). Grabbed urban landscapes: Socio-spatial tensions in green infrastructure planning in Medellín: Grabbed urban landscapes. *International Journal of Urban and Regional Research*. https://doi.org/10.1111/1468-2427.12725

Barthel, S., Folke, C., & Colding, J. (2010). Social–ecological memory in urban gardens—Retaining the capacity for management of ecosystem services. *Global Environmental Change, 20*(2), 255–265. https://doi.org/10.1016/j.gloenvcha.2010.01.001

Bell-Williams, R., Irvine, K. N., Reeves, A., & Warber, S. (2021). Digging deeper: Gardening as a way to develop non-human relationships through connection with Nature. *European Journal of Ecopsychology, 7*, 1–18.

Benjumea, D. M. (2020). *Post-disaster reconstruction in Medellin informal settlements and the associated resilience dimensions of place-remaking*. Doctoral thesis, The University of Nottingham. http://eprints.nottingham.ac.uk/id/eprint/59367

Benjumea, D., Kato, Y., & Hua, K. (2022). Participatory approaches to enact meaningful interconnectedness with the natural environment: A case study in Singapore. *Cities and Health*. https://doi.org/10.1080/23748834.2022.2124012

Benjumea, D., & Olszewska-Guizzo, A. (2021). Coping with extreme circumstances through community-led local nature interventions: A science-based policy analysis. In F. W. Gatzweiler (Ed.), *Urban health and wellbeing programme. Urban health and wellbeing (systems approaches)*. Springer. https://doi.org/10.1007/978-981-33-6036-5_10

Camps-Calvet, M., Langemeyer, J., Calvet-Mir, L., & Gómez-Baggethun, E. (2016). Assessment and valuation of ecosystem services provided by urban gardens in Barcelona, Spain: Insights for policy and planning. *Environmental Science and Policy, 62*, 14–23.

Daniels, B., Zaunbrecher, B. S., Paas, B., Ottermanns, R., Ziefle, M., & Roß-Nickoll, M. (2018). Assessment of urban green space structures and their quality from a multidimensional perspective. *Science of the Total Environment, 615*, 1364–1378. https://doi.org/10.1016/j.scitotenv.2017.09.167

DeKay, M. (1996). Systems thinking as the basis for an ecological design education. In *Proceedings of the National Passive Solar Conference* (Vol. 21, pp. 366–373). American Solar Energy Society Inc.

DelSesto, M. (2020). People–plant interactions and the ecological self. *Plants, People, Planet, 2*(3), 201–211. https://doi.org/10.1002/ppp3.10087

Descola, P. (1996). Constructing natures: Symbolic ecology and social practice. In Descola, P., & Pálsson, G. (Eds.), *Nature and society: Anthropological perspectives* (pp. 83–102). London: Routledge.

Dooling, S. (2009). Ecological gentrification: A research agenda exploring justice in the city. *International Journal of Urban and Regional Research, 33*, 621–639.

Echeverry-Tamayo, J. D. (2019). Medellín: una ciudad de encrucijadas. Pobreza, modelo de ciudad y cambio social en el proyecto Cinturón Verde. *Territorios*, (40), 273–289. https://doi.org/10.12804/revistas.urosario.edu.co/territorios/a.5535

Elton, S. (2021). Growing methods developing a methodology for identifying plant agency and vegetal politics in the city. *Environmental Humanities, 13*(1), 93–112. https://doi.org/10.1215/22011919-8867219

Folke, C., Colding, J., & Berkes, F. (2003). Synthesis: Building resilience and adaptive capacity in social-ecological systems. In F. Berkes, J. Colding, & C. Folke (Eds.), *Navigating social-ecological systems: Building resilience for complexity and change*. Cambridge University Press.

Frank, B., Delano, D., & Caniglia, B. (2017). Urban systems: A socio-ecological system perspective. *Sociology International Journal, 1*(1), 1–9. https://doi.org/10.15406/sij.2017.01.00001. Retrieved October 20, 2017.

Franz, T. (2017). Urban governance and economic development in Medellín: An "urban miracle"? *Latin American Perspectives, 44*, 52–70.

Gibson, J. J. (1979). *The ecological approach to visual perception*. Psychology Press. https://doi.org/10.4324/9781315740218

Gould, K. A., & Lewis, T. L. (2017). *Green gentrification: Urban sustainability and the struggle for environmental justice*. Routledge.

Hunter, R. F., Rodgers, S. E., Hilton, J., et al. (2022). Groundswell: Community-engaged and data-informed systems transformation of urban green and blue space for population health—A new initiative [version 1; peer review: awaiting peer review]. *Wellcome Open Research 7*, 237. https://doi.org/10.12688/wellcomeopenres.18175.1

Inglod, T. (1996). *Nature and society anthropological perspectives*. In P. Descola & G. Pálsson (Eds.). Routledge.

Irvine, K. N., Hoesly, D., Bell-Williams, R., & Warber, S. L. (2019). Biodiversity and spiritual well-being. In M. R. Marselle, J. Stadler, H. Korn, K. N. Irvine, & A. Bonn (Eds.), *Biodiversity and health in the face of climate change* (pp. 213–247). Springer International Publishing. https://doi.org/10.1007/978-3-030-02318-8_10

Ives, C. D., Abson, D. J., von Wehrden, H., Dorninger, C., Klaniecki, K., & Fischer, J. (2018). Reconnecting with nature for sustainability. *Sustainability Science, 13*(5), 1389–1397. https://doi.org/10.1007/s11625-018-0542-9

Kohn, E. (2013). *How forests think: Toward an anthropology beyond the human*. University of California Press. https://doi.org/10.1525/california/9780520276109.001.0001

Krasny, M. E., & Tidball, K. G. (2009). Community gardens as contexts for science, stewardship, and civic action learning. *Cities and the Environment, 2*, 1–18.

Lach, N., McDonald, S., Coleman, S., Touchie, M., Robinson, J., Morgan, G., Poland, B., & Jakubiec, A. (2022). Community wellbeing in the built environment: Towards a relational building assessment. *Cities & Health, 0*(0), 1–19. https://doi.org/10.1080/23748834.2022.2097827

Langemeyer, J., Camps-Calvet, M., Calvet-Mir, L., Barthel, S., & Gómez-Baggethun, E. (2018). Stewardship of urban ecosystem services: Understanding the value(s) of urban gardens in Barcelona. *Landscape and Urban Planning, 170*, 79–89. https://doi.org/10.1016/j.landurbplan.2017.09.013

Lawrence, A. (2009). The first cuckoo in winter, phenology, recording, credibility and meaning in Britain. *Global Environmental Change, 19*, 173–179.

Lee Kuan Yew World City Prize. (2016). *World prize city of Medellin*. Retrieved September 10, 2022, from https://www.leekuanyewworldcityprize.gov.sg/medellin/

Lumber, R., Richardson, M., & Sheffield, D. (2017). Beyond knowing nature: Contact, emotion, compassion, meaning, and beauty are pathways to nature connection. *PLoS ONE, 12*(5), e0177186. https://doi.org/10.1371/journal.pone.0177186

Marder, M. (2012). Resist like a plant! On the vegetable life of political movements. *Peace Studies Journal, 5*(1), 24–32.

Mayer, F. S., & Frantz, C. M. (2004). The connectedness to nature scale: A measure individual's feeling in community with nature. *Journal of Environmental Psychology, 24*(4), 503–515. https://doi.org/10.1016/j.jenvp.2004.10.001

McPhearson, T., Cook, E. M., Berbés-Blázquez, M., Cheng, C., Grimm, N. B., Andersson, E., Barbosa, O., Chandler, D. G., Chang, H., Chester, M. V., Childers, D. L., Elser, S. R., Frantzeskaki, N., Grabowski, Z., Groffman, P., Hale, R. L., Iwaniec, D. M., Kabisch, N., Kennedy, C., & Troxler, T. G. (2022). A social-ecological-technological systems framework for urban ecosystem services. *One Earth, 5*(5), 505–518. https://doi.org/10.1016/j.oneear.2022.04.007

OPPCM. (2016). *Observatorio de Politicas Publicas del Concejo de Medellin: El control Urbanistico en el Cinturón Verde*. Retrieved September 10, 2022, from http://oppcm.concejodemedellin.gov.co/sites/oppcm/files/2019-08/bordes-urbanisticos-2016.pdf

Resilience Office. (2016). *Resilient Medellin: A strategy for our future. Medellin Resiliente: Una Estrategia para el Futuro. The Rockefeller Foundation 100 resilient cities.*

Sotomayor, L. (2017). Dealing with dangerous spaces: The construction of urban policy in Medellín. *Latin American Perspectives, 44*, 71–90.

Sotomayor, L. (2021). *Medellín's integrated urban projects: A planning tool to reduce socio-spatial inequality*. Lincoln Institute of Land Policy. Retrieved October 28, 2022, from https://www.jstor.org/stable/resrep43360

Urban Land Institute. (2013). *Medellin voted city of the year*. ULI Americas. Retrieved September 10, 2022, from https://americas.uli.org/medellin-named-most-innovative-city/

Varley, A. (2013). Postcolonialising informality? *Environment and Planning D: Society and Space, 31*, 4–22.

World Bank. (2022). *Urban population—Latin America & Caribbean*. Retrieved September 28, 2022, from https://data.worldbank.org/indicator/SP.URB.TOTL?locations=ZJ

Chapter 9
Cultural and Natural Heritage, Synergy in Regeneration of Tropical Cities: Laguna del Carpintero Wetland and Its Surroundings, Gulf of Mexico

Miguel Ángel Bartorila

Abstract In Latin American cities, with their incomplete urban development processes and marginalised natural spaces, adopt regeneration practices to recover and improve the existing spatial, social, and ecological dimensions. Since it is assumed that natural ecosystems are a key element in sustainable urban regeneration projects, this chapter aims to explore the synergy between natural and urban processes in urban renovation projects through the integration of wetland areas, particularly mangrove areas. This text dissects the Conservation and Revitalisation Master Plan for the Laguna del Carpintero wetland and its surroundings located in the port of Tampico, Mexico. The project focuses on the integration of urban biodiversity around open spaces. The author introduces the innovative methodological aspects adopted in the project, which is located in a tropical city context. It conceptually develops the essential material components for understanding and formulating projects for Sustainable Urban Regeneration, including ecological succession. Therefore, mutual adaptation between urban and natural components associates ecological restoration with public space and built heritage in the project logic.

1 Introduction

In Latin America social exclusion and environmental damage have to do, besides urban sprawl, with the encapsulation of homogeneous spaces without common areas of sociability and no positive contact with natural spaces. The urbanisation global process, from which Mexican cities cannot escape, has, as a result, dysfunctional dispersed cities (Arcia & Silva, 2019), which have to be thought from the point of view of revitalization (Forsyth et al., 2016). There is no social or economic exchange with no good neighbourliness; there is no welfare with no coexistence with nature.

M. Á. Bartorila (✉)
School of Engineering, Autonomous University of Queretaro, Santiago de Querétaro, Mexico
e-mail: miguel.bartorila@uaq.mx

E. Mangi et al. (eds.), *Innovative Public Participation Practices for Sustainable Urban Regeneration*, Urban Sustainability, https://doi.org/10.1007/978-981-99-9595-0_9

The incomplete urban process is a challenge for urban policies in Latin America, where the inhabitants, considering possible scenarios, could benefit from integrating the natural component. Natural spaces, a physical expression of biodiversity, apart from having intrinsic value and provide ecosystem services, are an invaluable resource to reconvert the actual inefficient territorial occupation model into a resilient and sustainable city. Pauchard and Barbosa (2013: 590) about Latin American cities point out that "*…there is an urgent need to understand how biodiversity and ecosystem services interplay in and around cities*". The possibilities for natural regeneration and available voids are a feature of cities with incomplete processes (Andrade et al., 2013; Herzo & Finotti, 2013).

The industrial city model linked to a productive economy with serious environmental consequences is over—it has an impact on land and energy consumption and weakens the social fabric. When considering the paradigmatic case of tropical cities, the regeneration of natural ecosystems is a key element. Tropical cities, unlike warm boreal cities, have specific characteristics. Biodiversity is one of them. Lugo (2014) points out that in the tropics the interaction among people, the green cover and institutions are a challenge to scientifically study. This text explains the proposal Revitalization and Conservation Master Plan for Laguna del Carpintero wetland and its surroundings (RCMPLCWS). Financed by the Mexican federal government and set in a tropical environment, the project aims at solving housing problems from the urban revitalization point of view. The project developed in the academic field and was later joined and supported by other government institutions and civil associations.

Among the intrinsic values that biodiversity has in tropical cities, Muñoz-Erickson et al. (2014) point out the importance of the green–blue cover, highlighting the predominant role of the natural component in urban planning. The first hypothesis in this chapter indicates that natural ecosystems, particularly those in tropical incomplete Latin American cities, are an accessible and essential resource in sustainable urban regeneration projects.

The case of the Mexican cities has followed an unsustainable and uncontrolled urban growth pattern, negatively affecting natural areas, and substantially increasing the cost to provide public services (ONU-Habitat, 2019). This pattern of urban expansion has multiple causes, including social inequality and an opaque land market (Arcia & Silva, 2019). About Mexican cities Pauchard and Barbosa (2013) state that it is urgent to bridge the gap between ecological research findings and decision-making related to urban development. They also state that it is urgent to get citizens involved into supporting urban development based on ecosystem. The authors point out that "*Urban areas and biodiversity need not be mutually exclusive, and cities can –and should– promote inclusiveness of nature and wildlife in the urban landscape, rather than maintaining barriers*" (p. 603).

Mexican cities need to check the extensive urban growth territorial model. For Bartorila and Díaz Montemayor "*the territory is the foundation of urban and environmental regeneration*" (2022: 276). The conservation and integration of natural spaces in the Mexican cities, according to the authors, promote an efficient urban artifact, transforming its marginal spaces into urban ecotones. The urban ecotone both revalues local geographies and cultures, and invigorates identities and a social sense of belonging. Consequently, reurbanisation projects require more sensitivity toward the natural and social environment, more in-depth discussion and formulation. Therefore, the transformation of existing urban areas is a process where welfare public policies and existing social fabric with relicts of natural ecosystems are the seed money for a sustainable urban regeneration. In Mexico, on the other hand, as regard public policymaking, some challenges remain: lack of alignment and asymmetries among jurisdictions and practices of urban projects with low participation.

The second hypothesis is to redefine strong foundations for the conceptual building of an urban regeneration project, integrating natural to urban processes with a view to designing a master plan. It is necessary to innovate in the way proposals are drawn up and developed with the material elements. A new more compact and sustainable city model offers possibilities to solve problems using more open spaces to coexistence and more hybrid housing for prosperity.

The urban revitalization project is developed from the sustainable model but from an ecological perspective, which intends to make the best of the urban artefact and to reconfigure the relation with nature as well as to enhance social cohesion and stability (Rueda-Palenzuela, 2019). Thus, three arguments are proposed to generate urban revitalization strategies: first, enhance the way in which people interact with natural ecosystems, allowing for regeneration; the strategy is adaptation to the territory context. Second, overcome the gap between sprawl and compact city; the strategy is urban efficiency. Last, advertise the life-work link with employment rise and mixture of socio-economic levels to increase mutual interchange; the strategy is work-life stability. It comes down to a balance between conservation and development (Bartorila, 2020). The Urban Revitalization Scenario is tested by the articulation of three elements linked to each of the three strategies: Public Open Spaces, Vertical City and Economic Activities (Fig. 1).

Next, the text deals, in the first place, with common heritage, i.e. cultural and natural, as the main resource for sustainable urban regeneration, which is the sector

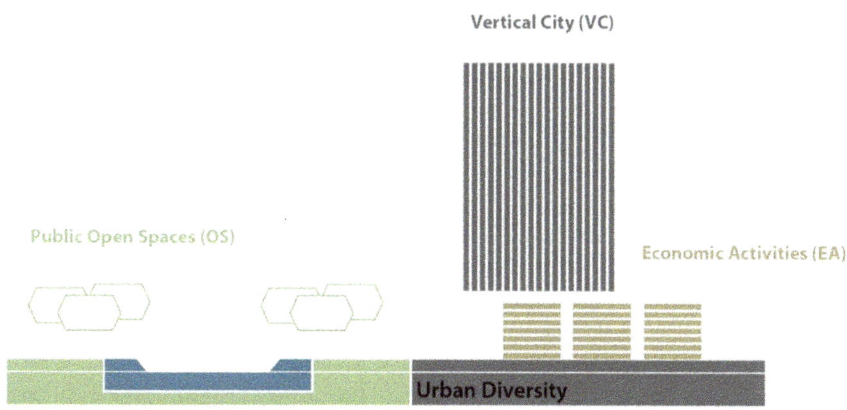

Fig. 1 Urban revitalization scenario diagram (*Source* Prepared by the author)

diagnosis. In the second place, cooperation between urban and natural processes is presented in developing a conservation and revitalization master plan.

2 Common Heritage, Main Resource to Sustainable Urban Regeneration

The pre-existing elements are a useful resource in the innovative practice of revaluing cultural and natural heritage. In order to turn the environmental and territorial dysfunctions in Mexican cities, an efficient, sustainable model of compact city is proposed—a model with a balanced approach on the vertical city, aware of the social reality and particularly of the ecological reality. In the case of the port of Tampico, there are cultural and natural resources highly recommendable for urban transformation: the modest architectural heritage and mangrove.

The Laguna del Carpintero wetland (LCw) environment, within the Tampico metropolitan area on the Gulf of Mexico, is a depressed and under-utilized area nearby the historical city centre but with high landscape value. It stands for the sprawling city model with weak ties with the natural surroundings. The lagoon, which is part of the Pánuco river wetland, is surrounded by vacant lots and low-density houses with huge services, buildings and infrastructure. The study sector, which has 563 ha, has natural space with great ecological value (192 ha); it is surrounded by neighbourhoods (371 ha) with enormous potential for higher urban densification: it only has 38,178 inhabitants and 8,507 houses.

The diagnostic made upon the Laguna del Carpintero area needs a multidisciplinary approach on the situation with a sustainable view (Bartorila & Rosas-Lusett, 2018). On the one hand, the historical background shows an architecturally typology-varied heritage catalogued by the Instituto Nacional de Antropología e Historia from Mexico. Besides, the study of both the modern and traditional collective habitat of the area recovers collective values which coincide with local ways of living, the house being considered modest architectural heritage. On the other hand, the hydrological analysis and the study of the wetland ecological valuation are complemented by urban sustainability rates, which have a more accurate dimension of the environmental and socio-ecological parameters. Mangrove areas, testimony to urban ecosystems, are a heritage which needs to be known and taken care of.

Tampico's surrounding wetland area has unique ecological qualities. The LCw is refuge for many flora and fauna species—mangrove and other 60 species as well as more than 41 animal species (Gutiérrez et al., 2018). Three mangrove sectors are affected by extensive urbanisation in Tampico (CONABIO, 2020; ONU-Hábitat, 2018). Mangrove areas provide environmental benefits through ecological services and their areas have been notably reduced around the world (Spalding et al., 2010). Their protection depends on the integration ways to the urban process. However, the logics, which have transformed the urban artefact, are above the fragile natural systems (Loredo-Cansino & Bartorila, 2021).

According to its ecological assessment, a vast area of the LCw falls short of its natural vocation, which manifests itself in the ecosystem fragmentation. The lagoon and its surrounding area can be typified as urban wetland. It has been surrounded and moulded by anthropogenic activity since more than 90 years through landfill work for land reclamation, therefore causing a reduction in the water body as water collector. One of the serious factors causing the wetland to reduce and its ecological function capacity to disappear is changes in land use. These changes were made with no previous assessment of flora and fauna diversity; let alone all ecological services this area has to offer to the whole ecosystem.

The manner in which the LCw has been handled, has caused ecosystems to fragment, consequently losing habitats and posing threats on biodiversity. The current situation, stated in the diagnostic, says that it offers alternatives of sustainable use for the wetland. The above-mentioned diagnosis uses ecosystem health indicators which associate flora and fauna.

The Bird Abundance rate (IAb), which is obtained comparing taxonomic lists from the LCw and from the Área de Importancia para la conservación de las Aves (Aica) No. 88 "Humedales del Sur de Tamaulipas y Norte de Veracruz", is 0.094, that is, LCw has almost 10% of all birds reported by the Aica, which amounts to 348 species. These figures show the importance which LCw has in the environment dynamics, in an area less than 1% from Aica No. 88 (Gutiérrez et al., 2018).

As regards the restoration of functioning habitats within cities, Handel et al. point out that "*Urban sites available for restoration ecology progress are usually small, surrounded by urban infrastructure, and isolated one from another*" (2013: 665). This fragmentation constrains the quality of natural communities. Therefore, one of the main strategies to preserve the wetland, according to habitat type and fragmentation causes, is to minimize the edge effect and heighten fragment connectivity. It is recommended to promote replanting and reforestation with native species through ecological restoration. Biological corridors allow organisms to move through and serve as refuge for species.

Taking into consideration the background and diagnosis from the wetland sector, urban redensification, transformation and conservation shall be proposed through a master plan. The project states that ecosystems and intraurban natural relicts can be configured as key components for regeneration in tropical cities. Natural heritage and urban opportunities associate to dig into urban revitalisation basic materials, assert local identity and respond to climate change, as it will be seen in the next section.

3 Cooperation Among Natural and Urban Processes in Building Regeneration Projects

The shift in conceptions of urbanism reflects a renewed centrality of the physical dimension of cities and territories in all of its ecological, social, and political complexity. Viganò states that "*Urbanism today is a field of innovation and a critical conceptual space where new issues and tools can be explored, many of which refer to the growing weight of ecological arguments and rationalities*" (2013: 407). Interactions between man and environment can be seen as co-evolutionary interrelation in which both sides bring about changes to each other continuously through mutual feedback as adaptive process (Gual & Norgaard, 2010). In this case, the RCMPLCWS seeks to achieve synergy among the different urban and natural processes in the material structure of urban sustainable regeneration. To the structural components of urban transformation, ecological succession is integrated as a process that allows rescuing and regenerating natural ecosystems in regression. The methodological contribution in the master plan conceptual development presents basic transformation components. Therefore, the ecosystem succession, as material component, articulates with property structure, public spaces, infrastructure, and architecture through co-operation.

The master plan summarises new knowledge resulting from coevolution, design and exploration practices on the porous city. The porous city illustrates the search of vital space as a plural and collective issue (Benjamin, 2007), and digs into ecological compatibility (Secchi, 2007). The master plan's conception and its configuration will be explained by means of division into in basic elements, description of two complementary territories and adaptation to territorial context.

Material Components for Porous City

There are four basic material components behind the plan, which were present in both the regeneration and on the critical insights for the transformation. The master plan proposal is divided in elements which summarise the structural forms from the 2060 scenario in contrast with the initial situation. The Solà-Morales (1997) hypothesis is further developed about the material components in urban growth models (land parcelling, urbanization and edification), with territory pre-existing natural elements and its regeneration dynamics (succession).

Succession is the process an ecosystem undergoes to develop and mature and to recover its space in urban areas. Conservation, maintenance or restauration of natural spaces are integrated to land parcelling, urbanisation and building. Thus, the territory with its natural component is a key element for ecological services, especially in tropical areas. Succession is, therefore, the available space for the ecosystem to regenerate. Land parcelling shows the process of limiting land ownership; urbanization is distribution of services and public space, and edification is construction of buildings. The knowledge derived from the proposal is $S + L + U + E$.

Both the initial configuration and the future reconfiguration summarise the four components. Thus, the understanding of project logics which defined master plan morphology explores a different interaction among the components (Fig. 2). This reconfiguration considers mangrove areas around the lagoon as key elements in sustainable urban transformation. The new composition revalues pre-existing natural territory to regenerate and places it in a position to interact with other three components: property structure, open urban spaces and new housing.

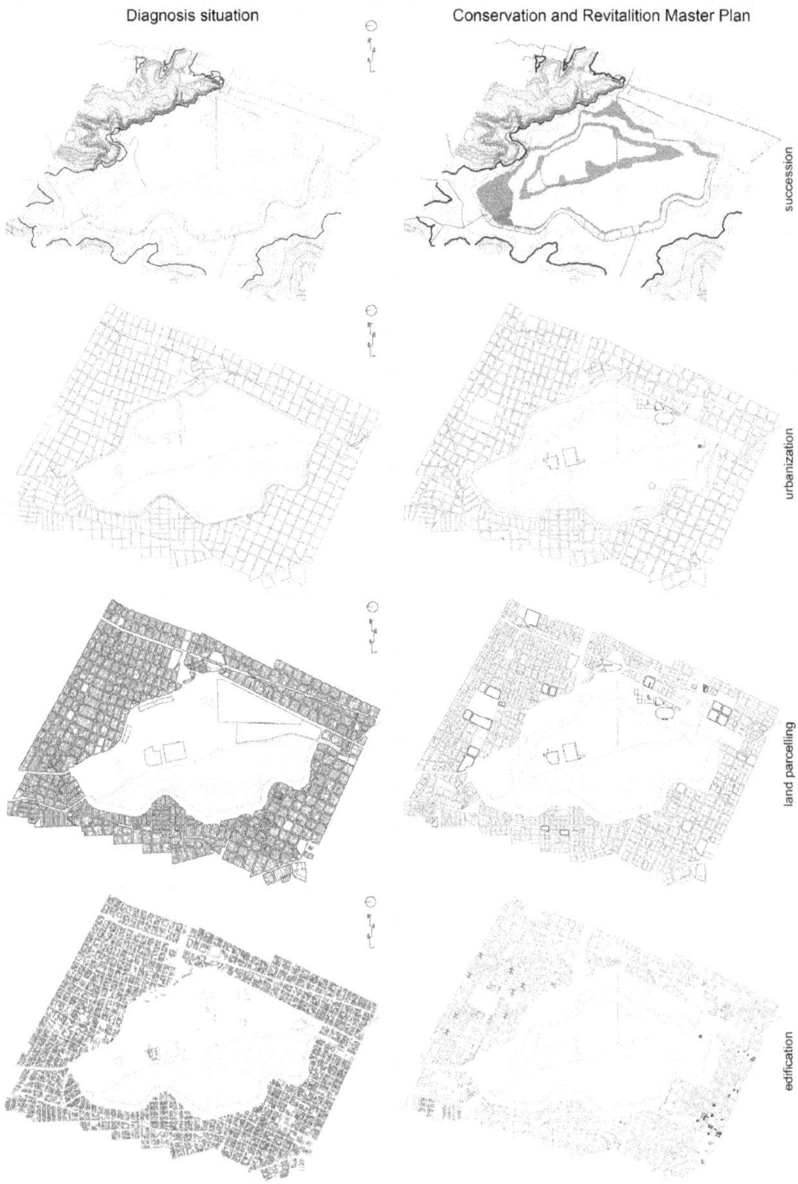

Fig. 2 Different interaction of the four components: succession, urbanization, land parcelling and edification (*Source* Prepared by the author)

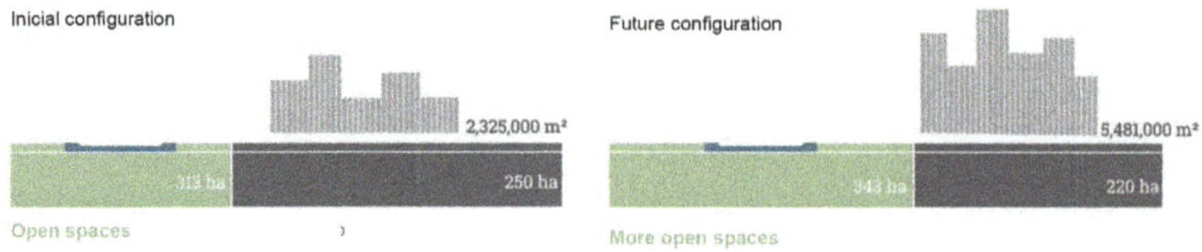

Fig. 3 Diagram of increase in open spaces and buildings (*Source* Prepared by the author)

Two Complementary Territories

In addition, the proposal for sustainable urban regeneration has two objectives: to recover the big natural public voids and to put forward measures as a means to make nearby neighbourhoods denser through new housing projects. The wetland ecological rescue includes restoring mangrove areas. The proposal also takes into account the local identity through revaluation of existing collective housing and built heritage; it seeks to integrate the conservation of natural spaces and development from the point of view of urban diversity.

Conservation and innovation, master lines for the urban revitalization plan, are deployed in two complementary levels. The first level is the urban natural space, the LCw, which is cleaned, emptied, regenerated, widened and spread through the urban context. The second level is a device which encompasses activities and inhabitants-nodes and passageways around the Laguna which revitalise the neighbourhoods making them denser. The result of more open space through transfer of land, and the increase in the building of housing and work spaces, is the synthesis of the sustainable urban model. Such a synthesis reveals clearly the urban compactness, which, in turn, shows the balance between all what is built, and the open public space people use to interact (Fig. 3). The scenario for 2060 foresees an increase in population, between 92,750 and 130,000 inhabitants; between 29,216 and 40,950 jobs in the sector and the construction of 32,605 new houses.

Adaptation to Territorial Context

Adaptation to territorial context through respectful transformation, which is the first strategy shared in the Introduction, is seen in the coexistence with water, biodiversity conservation and recovery of local identity. The Laguna del Carpintero surface area has reduced to half its original size causing flooding. The coexistence strategy with this body of water and its water basin promotes the system reconfiguration by increasing the water area, digging and rescuing pumping sumps and existing channels as well as increasing permeable areas (Fig. 4). Thus, permeable surfaces increase, vulnerable areas are defined and controlled and flooding risk decreases (Sánchez-Torres-Esqueda et al., 2018). Likewise, the importance of preserving bodies of water like rivers and lagoons reduces the heat island effect, the negative impact on comfort and demand for energy in cities (Rosas-Lusett, 2019).

In order to recover the wetland natural essence, conservation seeks to minimize the edge effect and heighten fragment connectivity (Gutiérrez et al., 2018). The

Fig. 4 Proposal for the revitalization and conservation master plan. General plan and sections (*Source* Prepared by the author)

integration with nature proposes the creation of Área Natural Protegida Humedal Laguna del Carpintero (ANP), a 147 ha area, to rescue and restore both wetland mangroves by tripling its area and other tropical ecosystems which are to restore continuity and to increase biodiversity. The area reconfiguration of LCw and its articulation with the urban area shows three sides: natural, recreational and cultural. As regard the natural side the ANP articulates trails and bird watching locations, which includes more than 91 ha water area and a 38 ha mangrove area (Table1). As wetland fringes are more sensitive places, alternative tracks to the inner park are proposed by land or by floating docks, avoiding concrete sheet piling and keeping natural soil to allow mangrove regeneration (Fig. 5). With the creation of ANP certain ecological services are strengthened like water purification and waste detoxification, climate control and climate change mitigation, cultural services and so on.

Finally, the plan urges to recover and restore 37.353 m^2 typology and scale-varied buildings, which add to citizen identity. The cultural heritage highlights historical and aesthetic values strengthening a community sense of belonging. A group of buildings remains protected: the modest architectural heritage, which are houses scattered in the neighbourhoods, as well as the catalogued patrimony, which is situated mainly in the historic centre. Likewise, 108 four-typology existing collective housing were

Table 1 Area reconfiguration of Laguna del Carpintero wetland

A. Laguna Carpintero Wetland natural spaces, coevolution and proposal			
	Original situation (ha)	Diagnosis situation (ha)	ANP proposal (ha)
Lagoon water body	168	73.35	91.43
Mangrove area	280	13.19	38.00
Other land ecosystems	0	33.49	32.38
Total			147.23

B. Area Reconfiguration Laguna Carpintero Wetland				
	Diagnosis situation		Proposed situation	
LCW Natural protected area	0	0%	147.23 ha	77%
Metropolitan park	10.53 ha	5.5%	32.38 ha	17%
Public buildings	34.47 ha	18%	11.22 ha	6%
Infrastructure and other areas	145 ha	76.5%	0	0%
Total	190.83	100%	190.83	100%

Source Prepared by the autor

Fig. 5 Perspective towards the center from the wetland (*Source* Prepared by the author)

selected, which intend to keep the experience of community living within the project boundaries.

Cooperation in the material and conceptual construction of regeneration leaves behind the mutual ignorance among natural processes and urbanism practices. The master plan proposes redevelopment and regeneration for the tropical city. It brings about a new urban centrality out of natural spaces with particular characteristics, like the LCw. The presence of water and natural relicts are the fundamental elements that address climate change.

4 Discussion and Implications

The presented work stresses two lines of discussion and its possible implications in redevelopment processes for tropical cities. First, it is the multidisciplinary approach in the making of the diagnosis and its interpretation. Second, it is proposed the natural and the urban simultaneously, highlighting contributions, synergies and cooperation.

The multidisciplinary element in the plan's study, diagnosis, conceptualisation and proposal, integrated the following disciplines: history, engineering, biology, economics, architecture, landscape design and urban planning. This process allowed building knowledge with contrasted, stronger arguments as well as more refined, concrete solutions. The complementary approaches from the multidisciplinary approach are to be seen, for instance, in the three concrete answers to climate change: reduction of flood risk by living together with the water landscape; conservation of biodiversity integrating nature and improvement of urban thermal comfort through reduction of the urban heat island effect.

Being able to rethink the urban natural space opened up a key discussion. The ecosystems specificity, in short, their revaluation, in the new urban pattern can be part of a new eco-environmental paradigm. To consider the LCw as an urban public space and a natural space simultaneously was the argument for design explorations, where public space typology concepts and ecological recovery were used, transcending the idea of green urban space. The coexistence between natural and urban processes simultaneously has been seen, for instance, in the interrelations among material components in configuring the project logics defining master plan.

From the project's perspective, urban transformation has to start from collective and multidisciplinary work and from co-operation among different public institutions even in the first stages of the regeneration process. Likewise, the point is made that producing knowledge to rethink how to make a city in the city is a high-value methodological contribution, which unveils a path to undertake the difficult task of regenerating the existing city.

5 Concluding Remarks

The ecological rescue presented in the chapter, revaluing voids and expanding ecosystems, is against urban space occupation practices. In other words, redefining voids, taking into account the natural qualities in the Conservation and Urban Revitalization Project for Laguna del Carpintero wetland and its surroundings, expands space in natural ecosystems and respects their shape. Therefore, an innovative practice in urban development is presented. Natural ecosystems in the city are important resources and significantly contribute to the urban balance as energy system dissipater. Hence, the necessary condition for ecological restoration to guarantee social and environmental benefits is to incorporate more space to the natural ecosystems respecting ecological succession.

The coexistence of different cultural perspectives as well as project practices with low participation are still challenges to strengthen sustainable urban regeneration projects. Biodiversity in Latin-American cities, represented by natural spaces as ecosystem goods and services, lives together with an incomplete urban developing process. The sprawling city model recommends revitalizing existing urban areas through public policies. Sustainable urban regeneration in Mexico is now beginning to open and is up against several challenges. In the case of Tampico and its shoreline wetland areas is revised from an ecological perspective. Also, an equation of the Laguna del Carpintero wetland revitalization and conservation is proposed. Beyond technical tools, building a regeneration plan is the right setting to integrate competing approaches and opinions not only from social groups but also from the silent voice of nature.

Acknowledgements The author would like to thank the Comisión Nacional de Vivienda (CONAVI) and the Consejo Nacional de Ciencia y Tecnología (CONACyT) in Mexico for funding the project "Revitalización urbana, oportunidad para redensificación. Ciudad vertical en corredores urbanos y espacios abiertos. Entorno de Laguna del Carpintero" (2015–2018), as well as the support from the Facultad de Arquitectura, Diseño y Urbanismo of the Autonomous University of Tamaulipas. Special thanks to Mario Camino for his support in style and translation.

References

Andrade, G. I., Remolina, F., & Wiesner, D. (2013). Assembling the pieces: A framework for the integration of multi-functional ecological main structure in the emerging urban region of Bogotá, Colombia. *Urban Ecosystems, 16*(4), 723–739. https://doi.org/10.1007/s11252-013-0292-5

Arcia, D., & Silva, E. (2019). Urban expansion in Mexico. In D. E. Diane, F. Vera, A. Seijas, & D. Arcia (Eds.), *City design, planning & policy innovations: The case of Hermosillo* (pp. 100–115). Harvard University Graduate School of Design. https://doi.org/10.18235/0001755

Bartorila, M. A. (2020). *Sostenibilidad y plusvalía. Proyecto de revitalización y conservación para el humedal laguna del carpintero y entorno.* Colofón. https://libros.uat.edu.mx/index.php/libros uat/catalog/view/250/208/688-1

Bartorila, M. A., & Díaz Montemayor, G. (2022). Urban and environmental regeneration in Mexican cities: A design framework. *Journal of Urbanism: International Research on Placemaking and Urban Sustainability, 15*(3), 261–281. https://doi.org/10.1080/17549175.2021.1893796

Bartorila, M. Á., & Rosas-Lusett, M. A. (Eds.). (2018). *Revitalización urbana: aportaciones para la ciudad vertical.* Colofón Ediciones Académicas.

Benjamin, W. (2007). Napoli. En *Immagini di città* (pp. 6–7). Gulio Einaudi.

CONABIO. (2020). *Portal de Geoinformación.* Sistema Nacional de Información Sobre Biodiversidad (SNIB), Comisión Nacional para el Conocimiento y Uso de la Biodiversidad. http://www.conabio.gob.mx/informacion/gis/

Forsyth, A., Brennan, C., Escobedo-Ruiz, N., & Scott, M. (2016). *Revitalizando Ciudades: Mejorando Viviendas y Barrios desde la Cuadra a la Metrópolis.* Harvard University Graduate School of Design. https://bit.ly/2VitqNV

Gual, M. A., & Norgaard, R. B. (2010). Bridging ecological and social systems coevolution: A review and proposal. *Ecological Economics, 69*(4), 707–717. https://doi.org/10.1016/j.ecolecon.2008.07.020

Gutiérrez, T., Ocón-Morales, S., & Balderrama-Alarcón, A. (2018). Valoración ecológica del humedal de la Laguna del Carpintero, Tamaulipas. En M. Bartorila & M. Rosas-Lusett (Eds.), *Revitalización Urbana, aportaciones para la ciudad vertical* (pp. 133–153). Colofón.

Handel, S. N., Saito, O., & Takeuchi, K. (2013). Restoration ecology in an urbanizing world. In T. Elmqvist, et al. (Eds.), *Urbanization, biodiversity and ecosystem services: Challenges and opportunities* (pp. 665–698). Springer. https://doi.org/10.1007/978-94-007-7088-1_31

Herzog, C. P., & Finotti, R. (2013). Local assessment of Rio de Janeiro City: Two case studies of urbanization trends and ecological impacts. In T. Elmqvist, et al. (Eds.), *Urbanization, biodiversity and ecosystem services: Challenges and opportunities* (pp. 609–628). Springer. https://doi.org/10.1007/978-94-007-7088-1_29

Loredo-Cansino, R., & Bartorila, M. Á. (2021). Higienismo, desarrollismo y naturaleza: lógicas de transformación entre humedales y ciudad en el Puerto de Tampico, México (1923–1964). Ciudad y Territorio. *Estudios Territoriales, LIII*(209), 805–820. https://doi.org/10.37230/CyTET.2021.209.12

Lugo, A. E. (2014). Tropical cities are diverse and deserve more social-ecological attention. *Ecology and Society, 19*(3), art24. https://doi.org/10.5751/ES-06618-190324

Muñoz-Erickson, T., Lugo, A., & Quintero, B. (2014). Emerging synthesis themes from the study of social-ecological systems of a tropical city. *Ecology and Society, 19*(3), 23. https://doi.org/10.5751/ES-06385-190323

ONU-Habitat. (2018). *Índice Básico De Las Ciudades Prósperas. Tampico.* Programa de las Naciones Unidas para los Asentamientos Humanos.

ONU-Habitat. (2019). *Reporte Nacional de Prosperidad Urbana en México. Índice de las Ciudades Prósperas (CPI).* Programa de las Naciones Unidas para los Asentamientos Humanos.

Pauchard, A., & Barbosa, O. (2013). Regional assessment of Latin America: Rapid urban development and social economic inequity threaten biodiversity hotspots. In T. Elmqvist, et al. (Eds.), *Urbanization, biodiversity and ecosystem services: Challenges and opportunities* (pp. 589–608). Springer. https://doi.org/10.1007/978-94-007-7088-1_28

Rosas-Lusett, M. A. (2019). Islas de calor en la zona conurbada del río Pánuco. *ACE: Architecture, City and Environment, 1*(39), 63–74. https://upcommons.upc.edu/bitstream/handle/2117/129914/5359-5242-1-PB.pdf?sequence=1&isAllowed=y

Rueda-Palenzuela, S. (2019). El Urbanismo Ecosistémico. *Ciudad y Territorio. Estudios Territoriales, LI*(202), 723–752. https://recyt.fecyt.es/index.php/CyTET/article/view/77733

Sánchez-Torres-Esqueda, G., Leal-Flores, A., Villeda-Oliveros, G. G., Ruiz-Del-Ángel, E., & Vargas-Castilleja, R., Ingenieros Sin Fronteras México, A. C. (2018). Informe Técnico de los Escurrimientos Medios Mensuales que se Presentan en las Micro Cuencas que Aportan Escurrimientos a la Laguna del Carpintero, Tampico, Tamaulipas. En M. A. Bartorila (Ed.), *Proyecto y estudio de viabilidad de revitalización en el interior de áreas urbanas: Caso humedal Laguna del Carpintero y su entorno* (pp. 162–184). CONAVI-CONACYT; Universidad Autónoma de Tamaulipas. https://bit.ly/2Kbab2n

Secchi, B. (2007). Rethinking and redesigning the urban landscape. *Places, 19*(1), 6–11. http://escholarship.org/uc/item/15q4w442

Solà-Morales, M. (1997). *Las formas de crecimiento urbano.* Editorial UPC.

Spalding, M. S., Kainuma, M., & Collins, L. (2010). *World atlas of mangroves.* Earthscan.

Viganò, P. (2013). Urbanism and ecological rationality. In S. Pickett, M. Cadenasso, & B. McGrath (Eds.), *Resilience in ecology and urban design. Future city* (Vol. 3). Springer. https://doi.org/10.1007/978-94-007-5341-9_25

GPSR Compliance
The European Union's (EU) General Product Safety Regulation (GPSR) is a set of rules that requires consumer products to be safe and our obligations to ensure this.

If you have any concerns about our products, you can contact us on

ProductSafety@springernature.com

In case Publisher is established outside the EU, the EU authorized representative is:

Springer Nature Customer Service Center GmbH
Europaplatz 3
69115 Heidelberg, Germany

www.ingramcontent.com/pod-product-compliance
Ingram Content Group UK Ltd.
Pitfield, Milton Keynes, MK11 3LW, UK
UKHW022210051026
15624UKWH00001B/96
* 9 7 8 9 8 1 9 9 9 5 9 4 3 *